PRAISE FOR LIFE UNDER GLASS

"*Life Under Glass* is a massively important and inspirational book about a great experiment that will be regarded as a cornerstone in the human quest to understand the Biosphere and ecology itself. Anyone who wants to understand what innovation actually is must read this book and whisper a hushed vote of thanks that people like this exist!"

         – Sir Tim Smit, Founder, Director, The Eden Project, Cornwall

"I am delighted to see a second edition of this important book that tells the true and, frankly, honest account of the Biosphere 2 experiment. It is important that the project has been so fully recorded here in what is also a most enjoyable read."

         – Sir Ghillean Prance FRS, VMH

"*Life Under Glass* is an honest, first-hand account of an innovative experiment. Biosphere 2 itself was visionary…the subjective experience in undertaking such an experiment makes enjoyable and inspirational reading for anyone eager to learn about exploring and discovering a new frontier where knowledge, skills, technology, and organization are key in managing self-organizing systems."

         – Dr. Jose Furtado, Centre for Environmental Policy,
         Imperial College, University of London

"I have known for five decades these deep, audacious, visionary nomads, walking the waves of the Heraclitus, creating Biosphere 2 to reflect our world, more real than true. *Life Under Glass* chronicles step-by-step a journey worthy of including in the Arabian Nights…braving the vivid unknown!"

         – Godfrey Reggio, documentary film director
         (*Koyaanisqatsi*, the *Qatsi* trilogy)

"*Life Under Glass* is a thrilling account of the daily life in Biosphere 2 and first and foremost a precious testimony on a unique experience that can teach humanity how to live in a small world, act as steward and feel interconnectedness. This book contains the keys to unlock the 21st century."

         – Jean-Pierre Goux, President of the Institute for Sustainable Futures,
         author, "*The Blue Century*," founder, OneHome

"*Life Under Glass* details an extraordinary scientific experiment, one in which a handful of idealistic citizen scientists, at considerable personal risk, volunteered to enter a closed system, Biosphere 2. The audacity of the effort brings to mind that famous quote of Teddy Roosevelt in which he hails not the critics, but those in the arena who strive valiantly, who spend themselves in a worthy cause, and who, if they fail, do so while daring greatly, their faces marred by dust and sweat and blood."

– **Professor Wade Davis**, BC Leadership Chair in Cultures and
Ecosystems at Risk, University of British Columbia, Vancouver

"*Life Under Glass* is a special present not only for me but for all the people who want to know, from real protagonists, the great history of Biosphere 2. The stories recounted here are extraordinary, beautiful, and dramatic at the same time. A must to read."

– **Antonino Saggio**, Professor of Architecture Sapienza
University of Rome

"*Life Under Glass* tells the story of an important experiment which has contributed to the void of neglecting ecological large-scale issues. These authors are explorers in the very best sense, storytellers at the finest. *Life Under Glass* hence allows a truly enjoyable read, accompanied by eye-openers directly from the authors' hearts."

– **Dr. Ralf Anken**, Head of the Department of
Gravitational Biology, German Aerospace Center, Cologne

"*Life Under Glass* is a great illustration of human creativity in extreme environments… the biospherian thinking process was an example of the emergence of a noosphere so that technics, or in my terms: art, science, and technology, reinforces life and life reinforces the arts, sciences, and technics in an evolutionary sustainable way."

– **Roger Malina**, ArtScience Research, Director ArtSciLab UTDallas,
Executive Editor Leonardo Publications, MIT Press.

"I am convinced that the epoch making experiment, Biosphere 2, will remain an indispensable topic of science education in high schools, colleges, and universities. The pioneering work of the biospherians may one day gain an additional importance for managing the life support systems of "Earth Observatories."

– **Professor Bernd Lötsch**, General Director (Emeritus)
Natural History Museum, Vienna, Austria

LIFE
UNDER
GLASS

SECOND EDITION

LIFE UNDER GLASS

Crucial Lessons in
Planetary Stewardship from
Two Years in Biosphere 2

Abigail Alling, Mark Nelson, and Sally Silverstone

FOREWORD BY SYLVIA A. EARLE

Foreword to the First Edition, Joseph P. Allen

SYNERGETICPRESS
regenerating people and planet

SANTA FE & LONDON

I LIVE IN A GLASS HOUSE

In the symphony of the biosphere
ecosystems do their riffs
the unseen more powerful than the obvious
we measure and observe;
the creation lives, Frankensystem or Alice in Ecoland
there's a momentum of its own
the people make decisions but hardly call every tune
species invade and conquer new lands
some vanish but their place is soon taken
this biosphere travels on its stomach
it's eat and be eaten
we work to elbow our way in
contemplating how many lobsters, lizards,
goats and people this world can support
we're leaner and meaner,
incredible shrinking biospherians
healthy and hungry
it gives an edge to our lives.

– **Mark Nelson**, 1992

Published by Synergetic Press 1 Bluebird Court, Santa Fe, NM 87508 & 24 Old Gloucester St.,
London, WC1N 3AL England

Library of Congress Cataloging-in-Publication Data

Names: Alling, Abigail, author. | Nelson, Mark, 1947- author. |
    Silverstone, Sally, author.
Title: Life under glass : crucial lessons in planetary stewardship from two
    years in Biosphere 2 / Mark Nelson, Abigail Alling, and Sally
    Silverstone ; foreword to second edition Sylvia A. Earle ; foreword to
    the first edition, Joseph P. Allen ; introduction to the second edition
    Mark Nelson, Abigail Alling, and Sally Silverstone.
Description: Second edition. | Santa Fe : Synergetic Press, [2020] |
    Includes index.
Identifiers: LCCN 2020004023 | ISBN 9781882428076 (paperback)
Subjects: LCSH: Biosphere 2 (Project) | Ecology projects. | Biotic
    communities--Experiments. | Ecology--Research.
Classification: LCC QH541.2 .A435 2020 | DDC 577.072--dc23
LC record available at https://lccn.loc.gov/2020004023

Cover and book design by Ann Lowe
Managing Editor Second Edition: Amanda Müller
Editor First Edition: Deborah Parrish Snyder
Cover photo by Gill C. Kenny
Printed by Versa Press, USA
This book was printed on Evergreen Skyland White Offset
Typeface: Gill Sans and Adobe Garamond Pro

TABLE OF CONTENTS

Biosphere 2, 1991.

"Yes!" was my immediate response to the invitation to be present and speak at the opening of Biosphere 2 after eight intrepid explorers had lived and worked within the confines of their glass-enclosed microcosm for two full years, September 26, 1991-September 26, 1993. Like many others, I had followed with fascinated interest what seemed to be an Arthur C. Clarke-like futuristic fantasy, but in fact involved real people living an otherworldly experience in real time. As a witness, as a scientist, and as one who had been part of a space simulation project more than twenty years earlier, I was intrigued and sometimes incredulous as the audacity of the Biosphere 2 vision became a successful reality. The Biosphere 2 team quickly demonstrated that one doesn't have to travel far to discover extraordinary new horizons.

Like Abigail (Gaie) Alling, co-author of this book, I am a marine biologist who literally becomes immersed in my research. I could not resist the opportunity in 1970 to live underwater for two weeks, leading a team of five women scientists and engineers during the NASA-US Navy-Smithsonian Institution-US Department of the Interior-sponsored Tektite II Project. Like the biospherians, those of us who lived as aquanauts isolated from direct contact with people on the outside were subjected to intense scrutiny by physicians, psychologists, and the public who were hoping to learn from the behavior of the ten teams who participated in the project, insights applicable to living in space and potentially, on the moon or other planets.

And, like the biospherians, we were keenly aware of the limits of our life support systems, from food and freshwater supply, to temperature, pressure, and especially the levels of oxygen and carbon dioxide in the air. But unlike them, we did not face the challenge of spending *two years* essentially self-contained, relying on living systems that had to be assiduously cared for to produce food and oxygen, absorb carbon dioxide, recycle wastes and otherwise provide for aesthetic and psychologically pleasing surroundings.

What I find most remarkable is that the Biosphere 2 system worked as well and lasted as long as it did before an outside source of supplemental oxygen was required. After all, it took four and a half billion years to develop Biosphere 1, Earth, as we know it: a living planet that is home to millions of species that together,

maintain sufficiently stable chemistry and temperature to persist as a place habit-
able for life in an otherwise extremely hostile universe.

Most surprising, Biosphere 2 worked despite a distinctly terrestrial bias, while
Earth is literally a "water planet." All life requires water and ninety-seven percent of
Earth's water is ocean. It is also where about ninety-seven percent of life exists and
where most of the oxygen in the atmosphere is generated, mostly by photosynthetic
plankton. The essential role of microbial life in shaping the chemistry of individual
organisms, of ecosystems, and ultimately of the entire planet has only recently begun
to come into focus. The ocean hosts a bountiful microbial minestrone of bacteria,
Archaea, and viruses that underpin the rest of life as we know it. Climate, weather,
temperature, and planetary chemistry are largely governed by the living sea.

Though small in size, and relatively limited in diversity compared to natural
coral reef systems, Biosphere 2's ocean appeared to be in good health when I was
privileged to dive into it, accompanying Gaie on a tour the day that the team
emerged. I was delighted to see not only healthy, living corals but also a number of
young coral colonies, evidence that spawning, settling, and growth had taken place
in their cocooned world. A common species of brown algae, *Dictyota*, was abundant,
and Gaie said she had to periodically "weed" the reef to prevent damaging overgrowth of the corals. It was exciting to discover a relatively rare kind of green algae
thriving amid various other algal species, invertebrates, and fish. I also noted when I
slipped underwater, that a number of confused cockroaches swam out of the crevices
of my borrowed equipment. Conscientiously, I rescued and returned them to the
shore. Although uninvited residents, they and a kind of ant clearly had made places
for themselves within the Biosphere's ecosystem.

Just as remarkable as the success attained by the biospherians at self-sufficiency
in a limited space with a limited ocean and limited sources of sustenance, was their
ability to maintain a civil, often congenial working relationship. It is a testament
to the creativity, discipline, dedication, intelligence, and overall good nature of the
participants that they successfully managed to not only survive but physically and
mentally thrive during their confined co-existence. Though bonded by common
purpose, shipmates at sea typically adhere to a certain mutually respected discipline
in order to maintain harmony. Similarly bonded by common purpose, occupants
of Biosphere 2's mini-world society had to cope with normal human differences,
preferences, habits, capabilities, quirks, and "personalities" of the others. This was
not a television program where individuals could get voted "off the island!"

As described in this thoughtful volume, the knowledge gained from the two-year Biosphere 2 odyssey has already inspired actions on many fronts, far beyond what may have been the original goals. Since the eight explorers completed their epic journeys within a confined space, interest has steadily grown in having humans as active participants in the exploration of the realms beyond Earth's atmosphere. The pioneering experiences recorded here serve as a vital baseline of evidence of what could be replicated in places currently unimagined. But far and away, the most important impact of Biosphere 2 may be the enhanced respect gained for the miraculous existence of Biosphere 1.

In closing, Bravo: Gaie Alling, Mark Nelson, and Sally Silverstone for sharing your knowledge and wisdom in this book, and for your continuing explorations, education, and sense of caring. Your legacy is real, your message clear: we must explore and care for this ocean-blessed planet as if our lives depend on it, because they do. A special salute, too, to the spirit of Biosphere 2 for inspiring the formation of the Biosphere Foundation as a means of fostering love and respect for the Earth among people globally.

– **Sylvia A. Earle**, oceanographer, author,
Explorer-in-Residence at the National Geographic Society

FOREWORD TO THE FIRST EDITION *by Joseph P. Allen*

EACH MORNING as I leave my home for work, I drive past the outdoor exhibit of a giant Saturn 5 rocket sitting beside the NASA/Johnson Space Center here in Houston. This dinosaur of launch vehicles and its smaller cousins form a kind of Jurassic Park celebrating the time when humans first left our home for bold but fleeting journeys to the moon. In the reddish, early morning glow of the August sun rising with a vengeance above Galveston Bay, the rockets take on a surrealistic look. Did we actually use nine of these behemoths to travel a quarter of a million miles out from Earth?

Such treks are difficult to imagine now that human activity on the space frontier has become so limited. NASA, a federal agency once known for its vision, imaginative engineering, and bold execution, nowadays devotes the bulk of its resources to planning projects which probably won't be implemented. This circumstance should not be surprising, since the demands made by an ever-changing, politically oriented space policy leave the beleaguered agency little extra resources to attempt something new. With such political impediments to overcome, will we ever again undertake space missions out to the moon and beyond? And if so, could we establish permanent human habitats out there?

Reflecting upon the heritage of these magnificent Saturn 5 rockets and fretting about the current uncertainties of our space program lead me to think more and more about a totally different kind of spaceship, Biosphere 2.

Strictly speaking, of course, Biosphere 2 is not a spaceship because it does not travel. Nonetheless, I am intrigued by the similarities of experiences of the Biosphere crew members with those of astronauts and cosmonauts aboard their ships in outer space. In spaceship terms, the people aboard this unique creation have been 'underway' for nearly two years now; and one month from the day I write this, they will open the seal which separates them from planet Earth and will exit from their world into an atmosphere as foreign to them as if they had been cruising out to Mars on the far side of the solar system. Does their speed increase now that they are homeward bound? Of course not, since Biosphere 2 has literally not budged one inch since the day the experiment began. Yet figuratively, at re-entry minus thirty days and counting, I picture the psychological calendar of the biospherians, having earlier crept forward with agonizing slowness when the Biosphere's food production waned and its oxygen level dropped, now flipping past pages with quickening speed.

I am also intrigued by this enormous experiment because, for the first time, it addresses the missing link of space colonies. By the missing link I mean the understanding of the closed sustainable ecologies needed for human habitation in space. All other technologies needed to live off Earth—rocket travel, for example—were proven during the golden years of the Space Age. But no understanding about closed ecologies was gained in those years because all space missions to date have relied on a rigid system of consumable stores: food, water, propellants, and so on carried according to a complicated flight plan and meted out piece by piece until they are exhausted about the time, one hopes, of Earth re-entry. Consequently, the single unanswered question still before us would-be space colonists is: can a closed ecological system be devised to be resilient enough to sustain human life for years at a time, yet of a dimension small enough (Earth-sized is clearly too big) to be constructed and maintained by normal human activity?

That this last and perhaps most complicated question would be explored first by a private venture rather than by a well-funded albeit ponderous government research establishment is, in my view, quite remarkable. Regardless of the outcome of these first experiments, the fact that steps toward understanding large biospheres have now been taken is to me both audacious and exciting.

As you read here about life under glass, you may find it difficult to imagine Biosphere 2 as another world. It is, after all, just *there*, separated from our own world only by a simple airlock. To go from one biosphere to another takes just minutes. On the other hand, space is surprisingly close by as well. The space shuttle, for example, travels out beyond the edge of Earth and into the vacuum of space in just over eight minutes, not much different than the time needed to pass through the two hatches and cross the anteroom of the Biosphere's airlock.

But major differences between Biosphere 2 and a spacecraft in orbit do exist— relative size and speed, for example. And there is always the non-trivial matter of the complex physics of rocket propulsion that launch a spacecraft into orbit, inherently dramatic in concept and still bold in execution even in this fourth decade of space travel.

The memory of a rocket launch is not something a person forgets. During powered flight aboard the space shuttle, the engines' roar pervades the crew quarters and the thrust of acceleration holds you against the launch seat at three times your normal weight. After the requisite velocity is achieved, the engines suddenly cut off leaving behind the eerie silence of coasting in unending Earth orbit. The three Gs of acceleration disappear as quickly as the sound. You unbuckle the safety harnesses holding you during launch and float from your seat to the nearest window. You, the space traveler, are now out of this world, privileged beyond all measure to gaze through a window that will forevermore change your perspective of both yourself and your home planet.

Watching Earth from orbit is breathtaking, awe-inspiring, tantalizing, and frightening—all rolled into one complex emotion continually evoked by the panorama before your eyes. Picture yourself floating at that window. Peering out, you watch the oceans and islands and landforms of Earth passing by your window at unimaginable speed. I want to write 'below your window', but in the weightless world of space you have no sense of 'up'. Thus there is no 'above' and no 'below' in orbital flight. You just float at the window and look out on the scene moving past at about five miles per second. Are you speeding by oceans and continents, or are you just hovering in a magical gondola and watching the world turn beside you?

The viewing angle of any part of Earth as seen from the spacecraft window is forever being changed by the relentless pace of orbital mechanics. You are constantly moving your head and hurriedly changing your body position, pressing always closer to the protective glass to catch the last glimpse of your favorite island cluster or to see precisely where your family and friends live there in Houston, marked by Galveston Bay and arrow-straight Interstate 10, the circle of the 610 Loop, and the familiar patterns of the runways of the city's airports. Within hours, the inside of all viewing ports of any spaceship are covered with forehead, cheek, and nose smudges which must repeatedly be wiped away.

Biosphere 2, of course, is not moving. Rather, it is firmly fixed in the hard-packed scrabble of the southwestern desert of North America, a beautiful part of Biosphere 1. The crew within the Biosphere views the outside scene, itself changing with the tempo of the seasons, at a leisurely pace. I suspect that most motion observed outside Biosphere 2 would be the parade of curious Earthlings peering through the glare of the glass walls to catch glimpses of the alien pioneers at work inside their independent ecology.

Along one wall of the living quarters, however, there is a special window through which inhabitants of both worlds can get a closer look at each other and engage in face-to-face conversation of a sort, the words carried by speakerphones, the images of the people slightly distorted by the internal reflection of the glass. I had the great pleasure of visiting with the biospherians through the conversation window, an occasion which started with greeting them by matching hands, a right hand flat against the left hand and left against right with the palms and fingers separated only by glass. Because of this rite of greeting, the conversation window is constantly covered with smudges of hand prints inside and out, quite unlike a spaceship. As far as I know, you will never find hand prints on the outside of a spaceship window.

The book these biospherians have given us here is also a special window of sorts. It gives us insight into the other world in which these unique pioneers have worked and lived for two years.

– **Joseph P. Allen**, former US astronaut
August 1993, Houston, Texas

"We invented Biosphere 2 not only for science, but also for beauty, adventure, and hope for all humanity – and for the Earth's biosphere itself. To teach human beings to see Biosphere 1 in a new way, this is the ultimate vision behind Biosphere 2. We have the ability to be a creative, cooperative agent with evolution. This is what I call victory."

– **John P. Allen**, Co-founder and Inventor of Biosphere 2
Research and Development, Space Biospheres Ventures (1984-1994)

"Space Biospheres Ventures represents a new approach to doing business. We are a private ecological research firm which has created one of the boldest research and development facilities of this century. We are also a profit-making venture. Biosphere 2 ushers forth technological development that is marketable and beneficial to the Earth. It responds to the current environmental crises by searching for real solutions, and stands as a vision of hope so that we as a species can move forward and leave our destructive ways behind us."

– **Margaret Augustine**, Co-architect and former CEO,
Space Biospheres Ventures (1984-1994)

"Biosphere 2 is the child of our Earth's biosphere, grown from the same flesh and genetic material, and born of the perspective gained with Apollo's distant images of the Earth. As with the Apollo images, Biosphere 2 allows us to see in one succinct view, a complete integrated system of life."

– **Edward P. Bass**, Co-founder of Biosphere 2,
Founding Trustee of the Philecology Trust

ACKNOWLEDGMENTS

BIOSPHERE 2 WAS THE RESULT OF A creative and dedicated network of scientists, managers, engineers, and architects who attempted the impossible—and succeeded. But a special word of recognition is due to three people: John Allen, for his vision, boundless energy, and practical savvy; Margaret Augustine, for her tenacity in guiding the dream towards its fulfillment; and Edward Bass, for his manifested commitment to the environment and the ecological technologies of the future.

The eight-member crew was only a part of this story; Biosphere 2 needs both a team on the outside as well as the inside. While the following pages give an insight into the challenges of living under glass, the entire biospherian crew is deeply appreciative of the extensive team on the outside whose support made our adventure possible. Without their dedication and vigilance, we could not have completed our task.

D. P. Snyder

Mission One biospherians from left to right: (Top) Taber MacCallum, Sally Silverstone, Linda Leigh, Mark Van Thillo (Laser), Mark Nelson, (Bottom) Jane Poynter, Abigail Alling, and Roy Walford.

LOOKING BACK ON LIVING IN BIOSPHERE 2

AND WHAT WAS LEARNED TO GUIDE US THROUGH
OUR CURRENT ECOLOGICAL CRISIS

"[Many people] believe that human society may successfully design nature to fit economic aspirations. What Biosphere 2 showed, in a short time, is the lesson that our global human society is learning more slowly with Biosphere 1, that humans have to fit their behavior into a closed ecosystem."

— **Professor Howard T. Odum, University of Florida**

IT SEEMS CRAZY TODAY, but in the 1980s the word 'biosphere' was not widely known; we often had to spell the word for people! 'Sustainable' was equally obscure. The reality that the global biosphere that evolved over billions of years was in fact the life support system for us humans, as well as for all other life on our planet, was not widely understood. Then suddenly, out of the blue, Biosphere 2 appeared housing many of the quintessential biomes of planet Earth: rainforest, savannah, desert, marsh/mangrove, and coral reef ocean. Not only that, but it was uncommon then, if not audacious, to have women as part of such an expedition team and certainly not in equal numbers with men, but our crew was composed of four men and four women. Everyone could relate to Biosphere 2 since it was a miniature version of their lives and world. Because it was so small, just over three acres (2.5 U.S. football fields or 1.5 soccer fields), it was easy to see how people, wilderness areas, farm, and technology are so closely interconnected.

Astronauts don't have to look over their shoulder; nobody is outside the spacecraft watching their every move. In Biosphere 2, even when we were slogging through a muddy rice paddy, cleaning the underwater ocean viewing windows, or pruning vines in the rainforest to let more sunlight in for shaded plants, there might have been a crowd looking in with their keenly peering faces pushed up against the glass that separated our world from theirs. Captivated by the drama of real-time science and exploration, they encouraged us with their positive smiles and were proof that what we were doing was both new and necessary. We also observed the outside world with new eyes: as we were getting thinner on our low-calorie diet, weaned away from fast food while living off our organic farm, people outside seemed to be getting larger! It was an opportunity for us to learn again about our own Earth's biosphere as we compared and contrasted the two while we learned from news reports that ecological problems around the world were getting worse. It was this comparison that helped fuel our resolve to successfully complete our two-year experiment in Biosphere 2 and provide some examples of how people take care of their biosphere when they realize it's what is keeping them alive.

The challenges we faced within Biosphere 2 have since moved from small environmental circles to the front pages of global news reports. Today, we continually hear that natural resources are finite and that even vast reservoirs, like our atmosphere or ocean, can be imperiled by human action. Our aim in Biosphere 2 was to find a way to meet our needs, protect our wilderness areas, and keep our soils, waters, and atmosphere healthy. To do this, we needed innovative technologies to recycle our water, grow food without using harmful chemicals while maintaining the soil's fertility, and when necessary, intervene to manage our atmosphere and preserve as much biodiversity as possible to enhance the system's capacity for adaptation and resilience. Our approach in the design of Biosphere 2, such as selecting non-polluting technologies to support ecology, directly relates to our Earth's biosphere because it too is essentially a closed system. That no one knew then if this was possible, made Biosphere 2 one of the most forward-thinking and daring experiments of its time.

Over the past two decades, we have continued to 'digest' and explore the unique journey we shared and have developed a greater appreciation of why Biosphere 2 was an important laboratory, as well as a life-changing experience for each of us. But Biosphere 2's accomplishments and relevance to our current ecological crises have never been fully appreciated. Hence, the importance of this expanded second edition of *Life Under Glass*, because the urgency of biospheric thinking and action can no longer be delayed without deepening the ecological catastrophe already underway.

BECOMING A BIOSPHERIAN

Since the three of us wrote *Life Under Glass* before we completed the two-year experiment, we decided to leave the book's original text largely untouched for its historical value. At the time, we didn't have the final data on most of the research projects underway, but we wrote with the immediacy and urgency of a still unfinished and dramatic experiment. We hope some of the rollercoaster ride of joys, frustrations and, above all, our total immersion and connection to a living world, will touch your hearts and minds. *Life Under Glass* is a record of how we started to learn to be biospherians, a term we called ourselves as a crew of eight. We realized in the early days of the experiment that our role was to learn from our biosphere, and adjust our actions accordingly to support its vitality, ecosystem functions, and overall health. We did not enter Biosphere 2 having a model of how a biosphere works; we started on day one to learn together.

The eight biospherians followed individual life paths to get to the project site in Oracle, Arizona at what was originally intended to be a quiet, although daring, research endeavor. We joined a small cadre of other biospherian candidates; we worked on the ecological and technical design; we traveled to exotic field sites collecting corals, plants, and animals; we operated prototype systems and even helped with construction while training in the skills required to live inside. The project's quick pace and high morale was familiar to NASA veterans of the Apollo Project who came to visit. Like them, we were not afraid of making mistakes. We were making history, attempting what was considered impossible. From the project launch in 1984, a research and development complex was soon built and operated to prepare for Biosphere 2. Breaking ground in 1987, hundreds of construction workers, glaziers, engineers, and specialists of all types worked at breakneck pace to complete that massive facility so that we could begin its initial shake-down mission in 1991.

Ultimately, eight of us were selected and coalesced to form a team for undertaking something that no other humans had ever attempted. We were going to be the first crew of the first human-made mini-biosphere for two years. It was a unique opportunity to study global ecology with us taking part as experimenters and part of a co-evolving sustainable living system. We had to carry out daily tasks that would enable our biomes to thrive, and us as well, for our breath, food, and health depended upon them. Although that was the goal of Biosphere 2's first closure experiment, everyone, including us, understood that it might not be achievable. At early Project Review Committee meetings, the sessions concluded with key managers and consultants enumerating lists of challenges, and sometimes even nightmare scenarios they could envision. Biosphere 2's initial closure experiment faced daunting odds and plenty of unknowns, but that only strength-ened our resolve to figure it out and do whatever it would take to keep our world healthy and biodiverse - and us inside! Ready for anything . . . we hoped!

A LABORATORY FOR THE STUDY OF GLOBAL ECOLOGY

Biosphere 2 was designed as a new kind of ecological laboratory, one in which the fundamental processes of life as well as that of an entire system could be investigated. To accomplish that, the facility needed to be energetically and informationally open (like our global biosphere) and virtually airtight (materially-closed) so that every-thing could be tracked with great precision. The structure was hailed an engineering marvel because its engineers succeeded in accomplishing daunting goals. Crucially, this included an unprecedented leak rate of less than 1% a month despite more than 20 miles of seams on the space frame roofs that braced against an outside Arizona climate that varied from subfreezing in the winter to well over 100 degrees in the blazing summertime. If Biosphere 2 hadn't been so tightly sealed, the decline in oxygen or the dramatic daily and seasonal fluxes in carbon dioxide couldn't have been detected and accurately examined.

Biosphere 2 became an ecological icon, recognizable around the world. Its biomes were tropical (it did not include temperate or polar regions) and its wilderness systems were chosen accordingly to include the biological, chemical, and physical require-ments that define each area. To achieve maximum diversity, both terrestrial and aquatic/marine mini-biomes were chosen along with differing ecosystems within each biome to enhance overall biodiversity. Project ecologists gathered about 3,800 species of plants and animals, plus uncountable microbes, fungi, and the other small-in-size but profoundly crucial microbiota in soils and waters that keep our global biosphere healthy and recycling. Because Biosphere 2 was vastly smaller than Earth, elements cycled faster. For example, there was a two to four day residence time of carbon dioxide in Biosphere 2's atmosphere compared to many years on Earth. John Allen, the inventor of Biosphere 2 and its executive director, presciently called the facility a "cyclotron for the life sciences." We were able to find things out rapidly, rather than over decades.

The technologies designed by our engineers also had to replace natural forces in order to supply much of what we take for granted since our planetary biosphere does it so reliably and for free; temperature control, rainfall, winds, tides, waves, and also supplement biogeochemical and nutrient cycles. There are no 'wastes' in a closed system. There is no dump or drain to dispose of unwanted substances; in fact, all "wastes" contain valuable resources. Thus, our engineering and ecological teams worked closely together to create new kinds of technical support systems that wouldn't harm life. For example, we made advances with innovative ways of purifying air using soil and plant biofiltration. In the ocean, we used a vacuum pump to make waves instead of centrifugal pumps that can decimate marine microfauna and flora. Wetlands function as the Earth's kidneys as they absorb and detoxify potential pollutants; so we worked with Dr. Billy Wolverton from NASA to design constructed wetlands to treat and recycle all of our human and domestic animal wastes produced inside Biosphere 2.

In a closed ecological system, our farm had to reach beyond organic standards to succeed in recycling all of its water and nutrients to grow enough food. Everything that came out of the soil had to be returned and we could not use any toxic products since they would build up in our soils and waters. The farm also had to be highly productive on a limited area of about half an acre. Though we did not succeed in growing all of our food during the first two years (we ate some seed stock during the second year inside), the 81% we did grow made Biosphere 2 one of the most productive farms in the world. Improvements made after our first closure mission, including using crops better adapted to the lowered light inside the facility, enabled the second crew to succeed in growing 100% of their food during their experiment in 1994.

The ocean system surpassed nearly all of our advisors' predictions; we built and sustained the largest coral reef ever created, let alone one in an artificial tank located over a thousand miles from the closest ocean basin, at 3,800 feet above sea level with temperate (not tropical) seasonal sunlight. When the ocean was resurveyed after the two-year experiment, 75% of the individual hard and soft corals had survived and some were reproducing. Even though the Biosphere 2 ocean had to deal with rising carbon dioxide, acidification, and low light levels, it persisted. Our mini-ocean taught us many things, but perhaps the single most important lesson was that it could adapt to new environments.

Many of our friends told us when we were building Biosphere 2, that it was 50 years ahead of its time. If we were to build it now, its purposes would be self-evident. In many ways, they were right.

Today, who can doubt the importance of developing highly productive ecologically-based farming systems to feed the world's growing population of around 8 billion while not polluting our rivers and oceans? Or applying methods to regenerate our farm and rangeland soils to hold more CO_2 rather than contribute to the climate crisis? Who can honestly dispute that deforestation and the demise of coral reefs and other ecosystems worldwide are fueling the frightening loss of species as well as diminishing our biosphere's ability to be resilient and adaptive? Why do we put up with pollution and the risks of exposure to untested chemicals when technology can and must be reinvented to not harm us, our ecosystems and our biosphere? Biosphere 2 successfully addressed all of these issues because it was a miniature replica of Earth's biosphere, as well as an ecologically engineered laboratory which made it possible to investigate how ecosystems function and interact.

Finally, and equally significant, the magnificent structure was inspiring. It was a celebration of architecture with stepped pyramid and barrel vault shapes, geodesic domes, a fjord-like ocean and expansive roofs to allow for the growth of trees during the project's intended 100-year lifetime. It showcased a new model that, by contrast, reveals the short-sightedness and self-destruction of our current assault on the global biosphere. We laughed when media accounts said we'd created an Eco-Disneyland. Our world wasn't make-believe; we were living and demonstrating that ecology and living ecologically is inspiring, gratifying, and meaningful. A good part of a generation grew up dreaming of becoming a biospherian with a vision of knowing that it is possible to live harmoniously with all other life. Despite the lack of the internet, when we exited the experiment in 1993, satellite TV coverage of our 're-entry' to Earth's biosphere reached close to a billion people around the world.

BIOSPHERE 2 PROVIDED INSIGHTS FOR LIVING IN SPACE

It's not easy to fully comprehend or study our Earth's biosphere because it is so vast. It is one thing to know that we live in a biosphere with finite resources and critical feedback loops requiring biodiverse systems, but it's another thing to learn with immediacy how it works or how our actions enhance or degrade its health. Thus, previous expeditions and space exploration were useful analogies for our planning. During the design phase, we met frequently to prepare for the two-year journey as if we were going to leave Earth. During the Biosphere 2 experiment, we observed ourselves and the challenges we encountered as precursors to what it would be like to live off this planet. We knew Biosphere 2 was to become our home for two years and that we would have to be resourceful, like we will need to be during future extended space exploration or in off-planet bases. In that respect, in addition to creating a laboratory for ecological experiments, Biosphere 2 also served as a test-bed to inform space design, generate baseline data about sustainable life support systems, and provide knowledge about group dynamics during long-term space missions.

During the two years, we came to realize that future space missions must include a wealth of living systems, for both physical and psychological health. All eight of us deeply understood that humans have evolved in and depend upon our biospheric life support system. Our ability to live in another world separated from Earth was only possible because Biosphere 2 contained a variety of Earth's ecological systems. The beauty of our ecosystems was nourishing; the life around us gave us joy, we were part of this new biosphere and evolved with it. That included developing and then fine-tuning its systems and our skills to provide for our needs; all of which will be required for people to live and thrive in space. The most important question for such space habitation is "what do we need to be OK once we leave this planet?" Answering this question produced years of discussions during the design phase that made Biosphere 2 a world where we could not just survive but live well.

A NEW EXPLORATORY SCIENCE AHEAD OF ITS TIME

The project presented a stark alternative to how "business as usual" threatens our Earth's biosphere because of our addictions to chemicals, plastics, fossil fuels, and technology that dominate our attention and divorce us from the natural world. Our wilderness biomes were off-limits to human expansion. They were valued for their own intrinsic beauty and uniqueness, and it was understood that biomes are the vital building blocks of a biosphere and critical for maintaining a healthy atmosphere and water cycles. Our rainforest was not cut down for short-term profits from timber and farming, nor did we slash and burn it or our savannah to create more farmland for growing food. Our mangrove marsh was not polluted with city and farming chemicals or other waste, nor was its water depleted for human use. Our coral reef was not overfished nor used as a dumping ground for trash or industrial and organic waste.

Our farm dramatically differed from industrial agriculture, one of the largest pol-
luters and sources of greenhouse gases. Biosphere 2's farm produced delicious organic
produce that was not suspect because we didn't (and couldn't, since we'd be harming
ourselves) use chemical fertilizers, pesticides, or herbicides. We recycled our wastes
and returned the nutrients to the farm soils rather than letting them pollute ground
and surface water and create marine dead zones. Our animals were champions in this
recycling system; they were both our companions and part of a successfully working
agricultural ecology.

Greenhouse gases and global climate change were not the front page issues
that they are today due to decades since of accelerating man-made impacts. Yet
back then, we were studying the cycles of carbon dioxide, methane, and nitrous
oxide in Biosphere 2 as a normal atmospheric aspect of any living system. As we
recount in the following chapters, one of the greatest challenges we faced was preventing a runaway rise in carbon dioxide from soil respiration and the imbalance
with photosynthesis as Biosphere 2 began with small plants and trees, our initially
"bonsai biosphere." During those first two years, our plants doubled their biomass
and a few years later, our soils reached more stable levels of organic matter. From
necessity and with ingenuity, we developed numerous ways to help control our
CO_2 production and demonstrate how humans can become involved in managing
the atmosphere and cooperate with natural systems, which have the capacity to
sequester (store) and recycle these gases.

Perhaps it was inevitable that Biosphere 2 would become controversial once
it surprisingly became headline news. To our dismay, even during our two years
inside, the facile and endlessly repeated media story was to recount the problems
and surprises of the first closure experiment and declare "Biosphere 2 is a failure."
Some even claimed Biosphere 2 was intended to be a substitute for taking care of
Earth's biosphere. On the contrary, Biosphere 2 showed us how irreplaceable our
global biosphere is and underlined the necessity for humanity to co-evolve with
it and become its stewards, not its destroyers. And with time, it became painfully
obvious to the eight of us that most media reporting missed a most crucial point: it
was above all else an *experiment*. We built Biosphere 2 not to demonstrate what we
knew but to find out what we didn't know, to learn from our mistakes and what the
facility, with its complex ecological systems, would teach us.

Because it was cutting edge science, it was bound to step on toes, but who
and what did it challenge? We inadvertently found ourselves in the middle of an

ongoing fight among scientists. Some small-scale reductionist scientists literally couldn't understand the systems level, holistic science of Biosphere 2. Arrogantly, they declared it not science at all! Worse, though, they perceived us as outsiders, mavericks, and not properly credentialed to undertake this project even though we had many world-class scientists, engineers, and institutions working with us to design and operate the facility. The fact that Biosphere 2 was privately funded and created outside the normal channels of government and university science grants, is the reason that the project moved so quickly. That independence allowed us to build a capable team and carry out the entire program which included the design, construction, and operation of Biosphere 2 for a price less than NASA would have spent just making the blueprints (as Joe Allen, a friend and NASA astronaut, told us).

The two-year Biosphere 2 experiment set world records for living inside a closed ecological system, eclipsing the six months that our Russian friends and collaborators had achieved in the most advanced experimental life support facility, Bios-3 in Siberia. That system imported some food and exported solid wastes. Biosphere 2 was the first closed ecological system that provided a complete balanced diet including animal products (meat, eggs and milk) and successfully treated and recycled all human wastes.

BIOSPHERE 2 FUNCTIONED AS A SEPARATE LIVING WORLD

What is astonishing, given the level of unknowns, is just how well Biosphere 2 did function and how successfully the ecological systems and technosphere meshed. There were surprises: the oxygen decline that no one predicted, or the fog desert that changed to more chaparral dominance; but perhaps the greatest surprise was how ecological zones in every biome remained largely intact. The coral reef, our greatest biomic challenge, which struggled with bleaching, coral disease, algae overgrowth, and lowered pH, surpassed all expectations. Only one hard coral species out of thirty-four was lost and 86% were considered to be in fair to excellent health at the end of the two-year experiment. The system as a whole functioned as a natural coral reef. The mangroves thrived, more than doubling in height, though understory species declined. The rainforest grew up rapidly and fulfilled the planned ecological succession: the fast-growing first canopy trees and ginger belt on all sides protected the more light-sensitive, mature rainforest species from the harsh Arizona sunlight. The ecological self-organization of these biomes provides lessons for ecological restoration of damaged ecosystems in the Caribbean, Amazon, Everglades and elsewhere around the world.

When we left, our world was lush, vibrant and remarkably diverse. The feared algal soup, mass extinctions, or merging of biomes into one weed-dominated eco-system never even remotely occurred. When we stepped out of Biosphere 2 and experienced such contrasting air, smells, sounds and the sight of the distant horizon in the Arizona Sonoran desert, we knew that the project had succeeded. We had indeed been living in a very different world and we palpably experienced the differences between the two with our first breaths of Earth air.

THE HUMAN DRAMA: LIVING WITH JUST SEVEN OTHER PEOPLE

Our greatest challenge was ourselves, humans, both within the experiment as well as outside. The 1991 to 1993 Biosphere 2 closure was a human isolation experiment that had never been done before. Boldly, the crew signed up for the two-year journey knowing that this would be a formidable feat of endurance and without a doubt we would face challenges, both personally and collectively, that were unpredictable. We had worked together prior to closure for many years, we knew each other well, but we also knew that the isolation and dramatic change in our day-to-day living would challenge us in ways we had never experienced before.

The experience of the small teams that spend winters in Antarctica, expedition teams, submarine crews as well as astronauts and cosmonauts confirmed that us-them divisions are common in exploration. Once a group of space station cosmonauts, annoyed at Mission Control, cut off radio contact for 24 hours; no questions were asked when they resumed talking to the people on the ground. Another cosmonaut described the extremely tight living conditions of the spacecraft he shared with two other people and dryly noted: "a perfect recipe for homicide!" Fortunately for us, our isolation experiment included wilderness areas, where we could drop out of sight of visitors and other crew alike, as well as spacious private bedrooms and public spaces.

Dr. Oleg Gazenko, Director of the Institute of BioMedical Problems in Moscow and confidant of generations of cosmonauts, was part of the Biosphere 2 Scientific Advisory Committee. He was most interested in the human dimension of Biosphere 2 and spoke with us during and after the experiment. He observed us closely and concluded that the biospherians had done far better than the cosmonauts in adjusting to our new environment, as evidenced by our feeling of freedom and ease with living inside. While some of us quickly adapted physically, feeling in sync with the rhythms of Biosphere 2, by the spring of the second year all eight of us experienced the shift as a team. We had become the Biosphere 2 team and our actions were relaxed, empowered, and coherent.

Dr. Gazenko also underscored the importance of the ethical standards of the Declaration of Helsinki, which requires that human research subjects be repeatedly given the right to leave an experiment at any time. While not a legally binding agreement, but an ethical one, it formed a fundamental experimental guideline for both Mission Control and all eight biospherians. Periodically, we came together to review our commitments individually and collectively. We all agreed that it was by our own choice that we took part in the Biosphere 2 experiment, but we retained the right to leave at any time. This also encouraged us to accept responsibility for difficult times since we all agreed to participate and work together.

Later, during the transition period after our closure was over, most of the biospherians reported that for weeks they felt more comfortable working inside the facility than outside. Psychological tests taken while we were still inside, in addition to private interviews held by the head of University of Arizona Medical School's psychiatry department, showed we were psychologically healthy, with nearly identical profiles amongst both men and women biospherians that closely matched those of other explorers, such as astronauts. One psychologist even told us if he was lost in the Amazon and needed help getting out safely, the eight of us would be his first choice!

Frictions and differences are inherent in any group; it's human nature. Just imagine being closed in for two years with only seven other members of your family, friends, or fellow co-workers. How well would you get along? The struggle over the future of Biosphere 2 that took place outside during the two years affected us inside, and probably added some fuel to the fire. Additionally, with the world's attention on us, just like other people in isolated, confined environments, there's a tendency to exaggerate these difficulties. But there were also plenty of fun and lighter periods, as well as tense periods throughout our two years. Remarkably, through thick and thin, we continued to work and play together. Any feast, morning with our few but precious coffee beans, or other celebration was too valuable to ruin with squabbles.

Biosphere 2 was our shared love and passion and all of the biospherians were totally dedicated to making our closure experiment a success and as scientifically fruitful as possible. Our world was also our lifeboat, so subconscious sabotage was unthinkable and never happened. We hope that also becomes a lesson for our situation now as an emerging global community faces very serious climate and resource challenges. It is imperative that people everywhere come to realize that Earth's biosphere is literally our shared home.

BIOSPHERE 2 NOW

When Space Biosphere Ventures changed hands in 1994 and the facility was given to Columbia University, it was altered from operating as a biospheric laboratory to one where the biomes were separated to create a greenhouse with controlled environments. The regenerative farm was dismantled and the space was used to study trees in different CO_2 environments. A crew no longer lived inside, the biospherian apartments were turned into offices and people and technicians entered to maintain the facility or collect data. The decline in the health and vitality of Biosphere 2 under this conventional scientific management illustrates that it takes both a detailed (reductionist) and a total system (holistic) approach to study a biosphere. The value of having live-in crews that could experience how each of our lives is dependent upon our biosphere was lost.

Research conducted when Biosphere 2 operated as a closed ecological system, from 1991-1994, produced a wealth of scientific data and publications. Unfortunately, much of the massive volumes of data collected from those first experiments are not available. Subsequent managers of the facility continued researching the ecosystems that were built, producing striking findings. Since the University of Arizona began managing the facility in 2007, they have carried forward the purposes of the project to serve as a center for research and education, in order to advance our understanding of the natural and man-made environment and catalyze interdisciplinary thinking and understanding about Earth and its future.Our vision was that Biosphere 2 would be the first of many biospheric laboratories that would be used for comparison with a range of different mini-biospheres to obtain critical knowledge about how life operates on planet Earth. It was named "2" to highlight that Biosphere 1 is our Earth's biosphere, the one we all share. As the first prototype, we expected it would give rise to Biospheres 3, 4, 5 . . . all of which would be built for biospheric research and to catalyze new technologies to restore our environment and generate environmentally-friendly economic technologies, such as new ways of cleaning our air and water and preventing pollution. To put that program in monetary perspective: Biosphere 2's estimated cost was approximately $200 million, which included land purchase, staff and consultant costs, design, construction, and operation of the experiment over the course of 10 years. This is equivalent to the cost of one modern military jet.

A NEW ERA

Geologists are now actively debating whether we are entering a new era due to the massive effects that a growing human population, industry and agriculture has had, and continues to have, on the biosphere. The new era would be called the Anthropocene because humans now drive many crucial trends on Earth. A few astute observers of Biosphere 2, like *Wired* magazine's Kevin Kelly, saw it as a forerunner of our modern biosphere where machines and human activities are interwoven with the natural world. To be successful in these early Anthropocene years, humanity has an overriding necessity to act thoughtfully and intelligently, becoming stewards of ourselves, and thus the health of our biosphere that sustains us. Biosphere 2 can be regarded as an Anthropocene experiment in its attempt to integrate appropriate technology with nature. Living in Biosphere 2 demanded that we stretch our human intelligence and capacities to consciously cooperate with our biosphere.

Vladimir Vernadsky, the great Russian biogeochemist who created our modern understanding of the biosphere, also thought evolution's new era would be the creation of a "noosphere" (a sphere of intelligence) in response to the geological power of human industry. He asserted that the noosphere will arise when scientific, technical, and biospheric intelligence harmonize so that technics reinforces life and life reinforces technics in an evolutionarily sustainable way. Our Russian colleagues, the leaders in closed ecological systems research, thought Biosphere 2 should have been named "Noosphere 1." We can now appreciate Biosphere 2 as an early laboratory experiment which foreshadowed a future when humans start to intelligently operate our "spaceship Earth." Then, we will start to transform our current downward trajectory into a regenerative, creative and evolving shared biospheric future.

WE ALL LIVE IN AND SHARE ONE BIOSPHERE

As we reflect on our years in Biosphere 2, it is clear that we were changed by it. Our bodily experience of being so closely interconnected with our biosphere was both remarkable and exhilarating. We understood on a profound level that our health and that of Biosphere 2 were the same. We were intensely aware that every action, everything we did, had immediate consequences. Our bodies understood and rejoiced in our cooperation with and dependence on all life. We had our responsibilities to work cooperatively with our living systems so as to maximize their well-being. We also understood the need to keep the support technologies functioning and upgrade them to be efficient, to accomplish as much research as possible in order to learn how our biosphere functioned. We also needed to become better farmers so we could eat more and alleviate the constant hunger accompanying our super-healthy but calorie-restricted diet. But the best, most fulfilling, and extraordinary experience was knowing, at a deep cellular level, that we were metabolically and consciously part of our living biosphere.

The fact that Biosphere 2 reached a worldwide audience, bringing into clear focus what a biosphere is, has been its lasting legacy. People were gripped as they followed the real-time drama we biospherians faced in keeping our world healthy. This helped puncture the fantasy that the environment is somehow outside of us and that humans are separate from nature. It was the beginning of a desperately needed planetary awakening. We all live in a biosphere and we are part of it! We are excited and hopeful now that this awakening is spreading so rapidly and widely around the planet.

We know that being chosen for Biosphere 2's first epic exploration was a gift for each of us. Even more importantly, we know that we are also biospherians of planet Earth, and so are you. Many have already changed their perspective and relationship to our biosphere which is overheating, burning, flooding, and whose overall health is rapidly declining. The warnings are clear: mass extinction of species, loss of much of the biomic diversity which powers our biosphere, the catastrophe-producing changes to our climate system, pollution and degradation of our soils and waters, the loss of natural regions and the beauty of Earth's biosphere. We are facing, for the first time, the real danger of so damaging Earth that our civilization and even our survival are threatened unless we reverse this accelerating destruction. Waking up to this crisis, and rising to the challenge of being responsible Earth biospherians, will enrich our lives. The crucial first steps are to realize that our biosphere is our home and our life-support system; and then to act, and act quickly, to restore it.

Our deepest desire is that Biosphere 2's legacy informs, inspires and offers insights into how to achieve this positive future.

– Abigail Alling, Mark Nelson, Sally Silverstone
February, 2020

At 8:15 AM on September 26, 1991, eight researchers entered Biosphere 2 to begin its first two-year mission of discovery.

CHAPTER I

The Adventure Begins

"At sunrise on Thursday, four men and four women will don red jumpsuits, share a hug with their friends in Mission Control and leave the world behind. If all goes well, they will leave the Earth behind for two years. The eight are not climbing aboard a space shuttle, although their language and nomenclature are deliberately evocative of the heyday of NASA. But they are embarking on an adventure that is in some ways bolder than the first manned space flight."

— *Los Angeles Times*, September 23, 1991

WE HAD TESTED THE AIRLOCK many times before in trial runs, but this time was for real. On a bright September morning, at about 8:15 AM, the eight of us—the first crew of biospherians—stepped into the airlock chamber to begin a journey some of us had been anticipating for seven long years. We had just waved farewell to hundreds of people gathered to see us off before we ducked into this unique compartment. The airlock was about the size of a cargo container, with gray, stainless steel walls, and two doors that had portholes like a ship. One door opened out into Earth and the opposite opened into the Biosphere.

As the metal door swung shut, helping hands tried to push down the large lever on the outside to seal it tight, our last help from outside hands for the next two years. Inside, we pushed down on our side of the lever, but the door didn't close. After a few moments, Mark Van Thillo called to those outside to step back, and with one decisive swing, brought the lever down to seal the door.

A few seconds later, we opened the inner door and entered Biosphere 2. Closure had been accomplished; our self-reliance had begun. The two-year challenge stretched out ahead of us, two years in which we hoped we would not need to go back through that door into the Arizona desert beyond it. With no one and nothing coming in, what was inside was all we had.

DAY OF CLOSURE

This historic day in our lives, the day of closure, was September 26, 1991. By 6:30 AM everyone—including the eight biospherians, the last-minute work crews, family, friends, and staff—had left the Biosphere and the doors were closed. The moment when an engineer closed the airlock door, our glass world became separate from Earth. No free flow of atmosphere, people, plants or animals, food or supplies would pass between Earth and the inside of the Biosphere again. All the preparations were over. From that moment on, the Biosphere became a distinct and separate entity, materially isolated from the rest of the universe except for energy and information flows. The physical boundaries were marked by the glass and steel space frames above, and the stainless steel liner below.

While we biospherians were taking part in the closing ceremonies in the plaza just outside the airlock, the Biosphere was already on its own. When we headed back towards the airlock at 8:15, we inserted the final, crucial ingredient into the experiment: the eight humans. Crucial because we were the ones who would either succeed or fail in making this extraordinary laboratory a working reality.

By the time the crew was sealed inside the Biosphere, along with our personal belongings and all the equipment for our mission, the day had already been crammed with activity and emotional intensity. We woke up at 3:30 AM while the Biosphere was still dark, long before the sun rose. We needed those early morning hours to prepare for satellite links with the East Coast morning TV shows airing three hours ahead of Arizona time. A handful of staff visitors, wanting one last night inside this world they helped build, lay curled up in blankets on the couches in the mezzanine. It was the last night that anyone but crew members would be allowed to sleep inside.

Those few hours before closure were filled with many 'last' things for us: last walks in the early morning desert air, last hugs from family and friends, last checklists, last treats. A jug of coffee brought by a friend was the last cup for those of us who hadn't already decided to wean ourselves from caffeine. Like many other luxuries, there would be precious little coffee inside the Biosphere; only having what beans we could harvest from the handful of young coffee tree saplings in our orchard and rainforest. And our teas were limited to herbal teas.

Opinions about how to prepare ourselves for such an unusual experiment were decidedly divided. Some crew members, such as Gaie Alling, Jane Poynter, and Mark Van Thillo (known as Laser), maintained that "the experiment begins when it begins," and they'd continue their normal patterns until then. Others, such as Mark Nelson and Linda Leigh (both coffee drinkers), decided to avoid the bodily shock they endured each time we started a week-long trial eating only what we'd have within the Biosphere 2 experimental diet, and so gradually cut out caffeine in advance.

The coffee jug exited with the visitors and the final clean-up crew. Along with the jug went the last packaged sugar and Styrofoam cups that we would see for a long time to come. Our very last luxury was a large breakfast of ham, eggs, and bread with butter; which we enjoyed in the peace and quiet of the Biosphere after the ceremonies, final goodbyes, and closure were over. From then on, all food would be grown, processed, and prepared by our own hands inside the Biosphere under the watchful eyes of Sally Silverstone, our co-captain, Agriculture and Food Systems Manager.

PACKING OUR BAGS

After years of intense concentration on the design and construction of an undertaking as ambitious as Biosphere 2, our own personal preparations seemed insignificant by comparison. So perhaps it's not surprising that some of us waited until the last two weeks before closure to get those needs in order. Many of us weren't even sure exactly what those needs would prove to be. How many pairs of socks, shoes, shorts, pants, shirts, and underwear will we need? What about clothes for special occasions? Would we even *have* special occasions to dress up for? Which books, tapes, CDs, photographs, paintings, stereo, TV, mementos, and other personal items should we bring in for our inner nourishment? This was a far cry from packing for a trip to Europe or a summer collecting expedition.

Sally lived out of a knapsack for years as she worked on various agricultural projects in India, Africa, and Puerto Rico, so material possessions were not a burden with which she had to deal. But others had more complicated situations. Roy Walford had cars, a house, and over a thousand experimental mice in his University of California at Los Angeles (UCLA) Medical School pathology laboratory. His cars he loaned to friends, his house he entrusted to his daughter, and his mice became the responsibility of his lab assistants.

Space Biospheres Ventures (SBV), the parent company that created the Biosphere, helped most of us store our clothes, furniture, and other belongings. But Mark Nelson brought all his clothes inside—from fur hats and heavy overcoats to dark suits

with black dress shoes to match. Roy brought his brightly colored *lungis* (one-piece
Indian cloth wraparounds for men). Jane brought a set of vividly colored wigs and
masks for parties and other lighthearted moments.

Jane and Gaie shopped for sneakers and blue jeans in the Tucson Mall. Jane
was the Manager of Field Crops and Animal Systems. Gaie (or Abigail) was the
Assistant Director of Research and Development and the Director of Marine
Ecological Systems for SBV, as well as Scientific Chief inside Biosphere 2. Her
responsibilities for the two-year experiment included not only the management
of the marine systems, but overall monitoring and management of the Biosphere
2 experiment, its research programs, and safety of its crew. Gaie earned a bache-
lor's degree in marine biology from Middlebury College and a master's degree in
Forestry and Environmental Studies from Yale, she had tracked whales and dolphins from Greenland to the Indian Ocean to Antarctica. Jane was a gardener,
trained in farm management in Australia at an Institute of Ecotechnics' project
with a stint of marine ecology on the Institute's ocean-going research vessel, the
Heraclitus. Both of them were accustomed to isolation in the outdoors and living
out of duffel bags. Even they miscalculated; they wound up with too many blue
jeans (twelve pairs in all) but not enough sneakers (six pairs). They figured that
their jeans would be the first things to wear out, because they'd be working daily
in the farm area, what we call the Intensive Agriculture Biome (IAB for short).
As it turned out, most of us wore out our work shoes before anything else. Jane
and Gaie's sneakers lost their heels and soles, which ended up in the scrap box.
Indeed, by the end of the two years, bare feet "outdoors" (outside our habitat
living area) became a common sight. In the habitat, it had already become a
custom, since we had to remove our dirty shoes to avoid tracking mud on the
carpeting. Sometimes going barefoot was a pleasure in our tropical world; but
it also had the practical purpose of minimizing wear-and-tear on the limited
supply of footwear.

Mark Nelson, knowing that half his time would be spent in manual labor in
the fields and wilderness, did his shopping in a quick trip to a couple of Tucson
thrift shops. A philosophy graduate from Dartmouth College, Mark, Chairman of
the Institute of Ecotechnics, had spent the last twenty years working on ecotechnic
agricultural, ranching, and ecological restoration projects in arid regions of the
United States and Australia. He was SBV's Director of Space and Environmental

Applications, and the Communications Officer for the crew. In the Biosphere he helped Linda Leigh, the Director of Terrestrial Ecological Systems for SBV, and manager of the systems inside, with her work in the terrestrial zones: the rainforest, savannah, and desert biomes. He was also in charge of the ecological waste recycling system and provided fodder for the domestic animals. Piles of new clothes seemed pointless; there was no one to impress, and clearly plenty of ways to get dirty. Laser (Mark Van Thillo) felt much the same way; work clothes would be just fine for him, too.

Jane had the idea to make a special box of new clothes for her to open on the first anniversary. Mark, although he relied on second-hand clothes, liked the idea and also prepared a box for the one-year anniversary of closure. This might not have been quite like a day's indulgence at the mall, but still there would be an infusion of something new.

We all had our Biosphere 2 jumpsuits, called by some our 'Star Trek suits.' The late Bill Travilla, one of Marilyn Monroe's clothes designers, was inspired by Biosphere 2 and offered to design something special for us. The results were these lightweight wool jumpsuits in fire-engine red and dark blue. We dutifully wore them for our official closing ceremonies and basically never wore them again. After losing a good bit of weight from the Biosphere 2 diet, they wouldn't have fit us again anyway.

Although Laser spent almost no time on his wardrobe, he meticulously prepared a collection of avant-garde music and a library of over three hundred books, rivaled only by Mark's collection, which he hoped to read in his free time. Laser was in charge of Quality Control for the construction of Biosphere 2; and once inside, his role was that of co-captain in charge of emergency systems, as well as the Manager of Technical Systems. A daunting task—he was responsible for the maintenance and operations of our very extensive infrastructure of machinery, which he had to keep running. Because he was intimately involved with the building of the system, he knew more than anyone else how everything was put together; therefore, along with his team on the outside, he was confident he could fix anything. If he and Mark ever managed to finish those five hundred volumes in their apartments with everything else going on, then they could have gone on to another thousand in the common library at the top of the sixty-five-foot tower in the middle of the habitat; housing a broad range of books in history, art, literature, architecture, ecology and other sciences, as well as philosophy.

Such was the intensity of the countdown to closure. And if anyone still suspects that we stashed bottles of champagne or jars of freeze-dried coffee, we can assure them that we did not! What we could grow in our fields was what we'd have. There was some food from crops grown in Biosphere 2 prior to closure, and all the fields had crops growing at various stages of maturity. Our goal was to see if we could produce all of our food during the two years, including replacing the original stored crops.

In addition to our personal wardrobes, we had piles of T-shirts, pants, skirts, and other clothes made of wool which were donated to us by The Wool Bureau, a company that promotes the use of natural wool products. The Wool Bureau liked the Biosphere 2 concept from early in the project, and believed that their products would work well in our tropical world. Their new design actually turned out to be recyclable, biodegradable, and comfortable even in our tropical climate. They also donated our carpets and the brightly colored fabrics on the walls throughout the human habitat area. The walls of the Biosphere 2 Command Room, our semi-circular office and computer/communications hub, were covered by a purple and gray wool fabric; the library walls an azure blue; and each bedroom was a different color of light brown, red, or blue.

Perfume and perfumed soaps or shampoos were not a good idea in the closed, recirculating atmosphere of the Biosphere. First, what they outgassed would confuse the detailed monitoring systems that tracked all the small amounts of trace gases that might be in the air. Moreover, in a closed system, compounds from non-biodegradable soaps or shampoos could easily accumulate to a toxic level. To put it bluntly, if we used a product that was not biodegradable, we'd be drinking it in our tea within a week. That excluded just about every readily available brand of hand soap, dish soap, detergent, and shampoo.

Sally tested many products, squarely facing the difficult task of satisfying both personal preferences and biological requirements. She once managed a hostel for mentally handicapped adults in her native England, and sometimes joked that it was this experience that made her able to survive as co-captain. Unflappably calm amidst the flare-ups of chaotic activity, Sally had been controller and General Manager of the architecture studio for Biosphere 2 during the entire design and construction stage, so settling the soap question was a minor matter for her. The biggest arguments centered around which soap to purchase, and needless to say Sally found a way to satisfy everyone. She included a granulated brand of oatmeal soap that hardly dissolved in water, as well as an oatmeal and winter-green soap that melted like soap when wet and could actually produce a lather. She stocked milk crates full of lotion, shampoo, conditioner, wintergreen and spearmint toothpaste, and natural sponges, all of which complied with the need to eliminate potentially dangerous gas emitters. Those of us who liked fragrances had to collect them ourselves from our herbs in the farm area and plants in the wilderness zones.

Quantities presented yet another problem: how much of the stuff would we need in two years? Some of us, like Linda Leigh and Gaie (who shared one bath-room), ordinarily used up to two containers of shampoo and conditioner (four ounces each) a week. It seemed unlikely that Roy, who was bald, needed any! Long-haired people needed twice as much shampoo as short-haired Sally or Mark.

There were many such questions to answer before the experiment could begin. What about feminine hygiene? Tampons and pads cannot be recycled, and if four women used these products for two years, the resulting garbage would be difficult to handle. We found an Ohio company called The Keeper that produced a reusable plastic cup designed to catch the menstrual flow. All that had to be done was wash it out. Toothbrushes also aroused controversy. Some dentists recommended twelve toothbrushes; others recommended four to six. In the end, a mix of electric and manual toothbrushes came inside.

Everyone had a thorough dental checkup before coming in, a requirement SBV was very firm about. Our healthcare specialists, Roy Walford, a then sixty-nine-year-old professor of pathology at UCLA Medical School who specialized in life extension; and Taber MacCallum, our youngest crewmember and Roy's assistant, had both taken the special US Navy course in emergency dentistry designed for use on ships far from land and without an onboard dentist. Their tales of old-fashioned tooth pulling and other dental tortures inspired what was probably some of the most conscientious tooth brushing in the history of dental care! Taber was a qualified deep-sea diver and Manager of the Analytical Labora-tory while inside Biosphere 2, but, so far as we know, he had never planned to practice dentistry.

Those with medical problems brought in special equipment. Mark, working on strengthening leg muscles following a knee injury a few years prior, split the cost of a piece of gym equipment for leg presses with Roy. Roy also brought a stationary bicycle, a rowing machine, weights, some isometric equipment, and a couple of braces for a neck problem. Mark included knee braces, heating pads, and electric massagers in his kit. At forty-four the second oldest member of our crew, Mark found that he was beginning to become farsighted. So he came in with two sets of reading glasses—one to start with, and one for his eye doctor's best guess as to what his eyesight might be after two years.

Since the Biosphere is covered with glass that excludes almost all ultraviolet light, there was no need for sunblock. In fact, we had to take vitamin D pills to make up for the lack of sunlight that the body normally uses to produce its own. We did bring sunglasses, however, because the light could be extremely bright inside; and of course, some used them to look cool.

Another useless item inside was money, although some of it floated in inadvertently in wallets or pockets. There was no money economy inside Biosphere 2. In time, other items came to be used for barter. Eventually, we bartered clothes, time, tools, but not food. Everyone ate their own as food was too precious to trade! Though, sometimes high-stake poker games amongst the crew were played for a handful of shelled peanuts.

ROOM FOR EIGHT

The eight of us were selected from an original group of fourteen biospherian candidates. We were a motley group: five Americans, two Brits, and a Belgian; four men and four women. When Mission One began, our ages ranged from twenty-seven (Taber) to sixty-seven (Roy). We were engineers, scientists, gardeners, and explorers.

Although many of us worked together for several years building the Biosphere, we still came in as eight individuals with unique histories. Gaie Alling, raised on the coasts of Maine and Georgia, experienced sailor and expedition chief, was one of the first marine ecologists to swim with sperm whales. Roy Walford, from California, was a world traveler and avid diver in addition to being the well-known author of *The 120 Year Diet* and a professor of pathology at UCLA who created four transgenic strains of mice and was one of the world's authorities on aging. Linda Leigh, raised in Wisconsin, a field ecologist as familiar with temperate ecology as she is with tropical, was our "wild woman naturalist" whose favorite comic book character is Swamp Thing. Laser Van Thillo from Antwerp, Belgium, was trained in industrial mechanics and engineering before heading off to work in India and Central America. Eventually he would board the research vessel *Heraclitus;* first as diver, and then as Chief Engineer. Jane Poynter's passion for gardening led her to range and crop studies in the remote Australian outback, where she began to work with domestic farm animals—unusual interests for someone born to the English upper classes. Mark Nelson, originally from Brooklyn, is a founder and Chairman of the Institute of Ecotechnics, a small think-tank organization devoted to harmonizing ecology and technology. One of his main concerns for many years has been to bring together space researchers, medical scientists, physical scientists, natural scientists, technologists, and agriculturists from all over the world to work together addressing environmental problems. Sally Silverstone was the resourceful English co-captain whose background, both academic and practical, in social and agricultural problems of underdeveloped countries probably helped us more than we may have ever predicted when the doors first shut behind us. Taber MacCallum, raised in New Mexico, had also been a world traveler, lived in Japan and Egypt, and had been the Chief Diver on the *RV Heraclitus* during expeditions to explore coral reefs and island cultures around Australia and the Red Sea.

Of course, aside from the differences in experience and professional background, we were all different in temperament, a very ordinary mix of 'night people' and 'day people,' extroverts and introverts, theorists and practitioners, doers and dreamers—all thrown in together to make ourselves into one unified crew with one unified goal. Each of us had our own unexpressed hopes and fears about the commitment we'd just undertaken. Perhaps some of us longed for the scientific joys to be found in this new world, or the special intimacy in living closely coupled to the life forms of an intriguingly complex, dynamic system. Some crewmembers visualized this world as an opportunity for personal transformation—but in what form, who could say?

There were the dark fears, too, which we could scarcely admit to ourselves, let alone to the others. Would Biosphere 2 go disastrously wrong? After all, we were stepping into unknown territory. Maybe there was a reason why no one had ever built a biosphere before, practical reasons that we didn't know about. The air could be poisoned by any one of the hundreds of trace gases; dangerous fungi might multiply and invade our bloodstreams. Plagues, locusts, fire, loneliness—we didn't know what really lay ahead.

Although SBV included redundancy and what were hoped were fail-safe back-ups in the technical designs, we all knew the infallible rule of machines: they break down. If the cooling systems failed, temperatures could rise above 150 degrees under the glass roof in just a few hours, a biospheric oven. And dozens, no hundreds, of other disasters could easily be imagined, including the crew being run ragged working from dawn to dusk to keep everything going. Murphy's Law could undoubtedly apply here as it does everywhere else: if nothing can go wrong, something will; if something goes wrong, everything will. This new world contained hundreds of machines, pumps, motors that could break down, tens of thousands of feet of cable and wiring ready to short-circuit. Were the eight of us heroes or fools for stepping in there?

We knew the world wanted to follow the experiment and how we eight were faring in our new world. The original idea that Biosphere 2 might be a quiet research and development project had evaporated when the US and international media began to run high-profile stories about what was happening in quiet, remote, Oracle, Arizona. None of us were prepared for an ever-increasing level of media attention. Thankfully, realizing we'd have to quickly learn the basics of dealing with the media, project management called in Carole Hemingway, of the Hemingway Media Group, to work with us. All eight of us had sessions with Carole and her partner Fred Harris, who became key parts of the Biosphere 2 media department throughout the year prior to closure. We learned how to effectively tell our stories and keep people informed about the progress of this "real time science" experiment which had captured the world's imagination.

When we crossed the threshold, we couldn't actually know if we would make it, although we were determined to do so and did not even admit the possibility of failure. From the Russian research in closed ecological systems, the less developed U.S. research, and SBV's own studies, we were confident that it was possible. But not only did we not have all the answers—we didn't even have all of the questions yet! This two-year experiment was the maiden voyage, the massive shakedown cruise for the most complex ecological experimental apparatus ever devised. If the system worked, Biosphere 2 would provide a powerful new experimental tool for the multidisciplinary science of biospherics, a controlled mesocosm in which to study global ecological processes both in detail and at a systems level, as a whole. The usually unmeasurable would become measurable.

Up with the crack of dawn, the first part of every morning is spent in the farm.

CHAPTER 2

A Day in the Life

"Biosphere 2 is a new kind of telescope which can be used to look at the Earth itself. We need to take the time to understand how to use it, to discover the kind of questions we should ask, and to scale up from there."

— **Dr. Christopher Langton**, Santa Fe Institute

June 2, 1992

INTERVIEWERS AND VISITORS always ask us what a typical day was like inside Biosphere 2. What did you *do* on February 10 or June 2? Did you mark your calendar with any special event that day? Or was it just another Monday or Tuesday?

Even if we didn't check our logbook or diaries, we could say that we spent Tuesday, June 2, 1992 in a way that was unquestionably different from anyone else on Earth. But to us it was a fairly typical day. Here is a look at what filled our hours that day; the explanations of events are brief, because many of these daily tasks were a continuous part of the much larger challenges and research studies which the remainder of this book will detail.

Dawn broke over the Santa Catalina Mountains at 5:35 AM that morning. No activities were scheduled at that hour, but the crew's early birds were already stirring. Linda got up to do her early morning check of the wilderness areas, including observations on how much had been eaten of the 'monkey chow' put out in bowls for the galagos (the small primates also known as bush babies) in the lowland rainforest. Sally made her early morning cup of mint tea and was already on her way to work in the vegetable patch of the agriculture system. Mark, who had also gotten up early, was hand-watering the supplemental crop boxes on the agriculture balcony, and cutting fresh mint and herbs for the kitchen. He would soon log into his email to review the weather report of the last twenty-four hours and record it in his notebook.

By 6:30, Gaie, the breakfast cook of the day, sounded the official wakeup call. She phoned each apartment, allowing the crew a half-hour to shower and dress before the hour of work that precedes breakfast.

There was one major difference between the bathrooms you're familiar with and the ones inside Biosphere 2: we had no toilet paper. There was no way that our recycling system could handle the amount of toilet paper that eight people would produce in two years. Instead, we used a water spray that hung next to the toilet. We found it in a plumbing catalogue designed for Saudi Arabian customers; the Arabs (as well as many other cultures) consider toilet paper far less effective for hygiene and have used water for the purpose for centuries.

Flushing the toilet, of course, didn't mean that 'it went somewhere' to be forgotten. All of the water that comes from the human habitat area—from toilets, showers, kitchens, laundries—went to the basement of the agricultural area and into the waste recycling system. Since we monitored the system every day, we could often tell if a faucet had been left open or a toilet was malfunctioning, because a suspiciously large amount of water would have entered the tanks.

After the wakeup call, Gaie headed for the kitchen to get breakfast going. She pre-heated the oven, boiled water for porridge and tea, and then stopped in at the command room to get the twenty-four-hour report on carbon dioxide to bring to the morning meeting.

By 7 AM, work had begun. Sally was milking the four she-goats, and Jane was feeding them. The buck and two bleating kids got their feed, too. Sally fed the chickens their portions of worms and azolla, a high-protein fern that grew on the surface of the water in the rice paddies. Later they'd bring the milk, plus eggs collected from the chickens, up to the breakfast table.

Jane checked the status of the irrigation tanks in the agriculture basement and the fields to ensure that the computer-programmed watering system had been triggered successfully. Sally began her daily collection of fresh vegetables, and since it was a Tuesday, she also brought up the next week's rations from the basement by elevator—burlap sacks of sweet potatoes, taro, flour, and beans. The supplies and the five-gallon buckets of vegetables Sally had picked had accumulated in the plaza outside the main double doors leading into the farm.

Linda, Taber, and Mark took pruning shears and sickles to the rainforest. Linda and Taber climbed into the space frame in the northeast corner to continue cutting back morning glory vines that were shading the trees. Mark was working by the *varzea*, the rainforest stream, cutting and bundling up morning glory vines that tangled the steep slopes of the banks and waded into the stream to cut and haul out their roots. By the end of the hour, Linda and Taber gathered twenty-five pounds of the most edible morning glory leaves and vines for the fodder storage bins in the animal bay, our enclosed barnyard area.

Laser logged on to his computer program in the command room to check the technical systems around the Biosphere. A special vibration analysis program gave early warnings of potential breakdowns. He studied the analysis report, completed the weekly maintenance report, then took a quick trip to the basement below the savannah to check on the tanks of water that had been condensed out of the air which passed through the wilderness biomes. Laser was in charge of the rain for the wilderness area (both terrestrial and marine) and had to mix the water which drained through each biome (its leachate water) with an appropriate amount of condensed water to make acceptable irrigation water for each area.

Meanwhile, Roy was completing the laboratory workups from the last set of biospherian medical checks. He was also in the process of conducting a stress hormone study, which requires the collection of daily urine samples. He added the fixative agent to the next day's collecting bottle and stored the previous day's samples in the freezers in the genetic and tissue culture laboratory on the mezzanine above the analytical laboratory.

By then, Gaie had finished fixing breakfast. Porridge was standard for every breakfast, and that day she made it with a mix of sorghum and wheat flour, sweetened with ripe bananas and papayas, topped with goat's milk. The rest of the menu depended on the cook's allotments of food. On special holidays and birthday mornings there might have been omelets or banana-filled crepes or even a cup of coffee from beans grown on one of the dozen young coffee trees. (The few beans we grew in the Biosphere were never enough for daily cups of coffee.) That morning, along with the porridge, Gaie served a carrot-cake loaf topped with an icing of milk, banana, and passion fruit; there was also a side dish of beans and sweet potatoes stir-fried with chilies.

At 8:00 AM, the kitchen chimes would finally sound on our two-way radios, and we assembled for breakfast. Our breakfast started off with the usual joking and social conversation, but it also served as our morning staff meeting, updating everyone on the progress of the experiment. Sally called the meeting to order and went over who was on the day's watch and who had cooking duty. She also asked Mark for the weather report, which included key environmental data: high and low temperatures in all the biomes for the previous day, the relative humidity in the agriculture area, outside temperatures (needed for gauging how to program our air handlers for cooling and heating), outside and inside total light received, and high and low carbon dioxide values at various sensors. Gaie then added the high and mid-point CO_2 for the previous day, which was followed by a discussion about various options to deal with CO_2 levels, tactics which had to conform with the SBV research strategy of minimizing the impact of elevated CO_2 on the overall system, and specifically to the ecology of each biome. Should temperatures be lowered in the biomes to lower soil respiration? What was the status of compost making which releases CO_2? When would the dormant desert and savannah receive their first activating 'monsoonal' rain, setting off an extra release of CO_2? Finally, Sally outlined the tasks for the morning agriculture crew, and each crew member outlined his or her day to make sure that all activities were coordinated.

After Sally adjourned the meeting, the 24 hour watch duties (similar to a ship's officer watch) was officially handed over from Gaie, who was on every Monday, to Mark, the Tuesday watch. Seven biospherians share the watch duties because Laser, as technical manager, has to be on back-up call for all of the others. If anything unusual had happened on Monday, or if there were any alarms during the day, Gaie would note them in the logbook. But this day had gone perfectly. Mark took over the watch and checked by radio with the Mission Control counterpart on the outside, who had also received the Mission Control watch handed over from the previous day's watch person.

With twenty minutes until the start of the morning work crews, we had some free time to catch up with messages on our computers, or the morning news on TV. Gaie did a quick cleanup in the kitchen, loaded the dishwasher, and turned it on. Then she brought a couple of jugs of mint tea to the plaza for morning break.

For all of us, the one-hour agriculture work began at 8:45, with five of us continuing on for another two hours. That day we weeded sweet potato, sorghum, and peanut plots in addition to routine agricultural duties. Laser, in charge of compost, began his hour by pouring several buckets of animal manure and crop residue into the hammer mill which shred the material into our compost machine and helped accelerate the decomposition process. His other responsibilities were to feed and water the worm-bed area in the agriculture basement. That morning he had brought a bucket of worms to the animal bay for the chickens. Mark's daily routines include harvesting a bucket of the azolla water fern and cutting fodder for the animals. That morning, he cut elephant grass that had been planted along every available walkway of the agriculture area. He also gathered a bucket of canna lilies that grew in the constructed wetland wastewater lagoons in the south basement which were also used as goat fodder.

While in the south basement where the light spilled through a span of glass, Mark checked the constructed wetland wastewater system, which consisted of three holding tanks that received all the wastewater from our habitat and another set of three for wastewater from the animal bay and laboratories. When filled, the tanks were closed and anaerobic bacteria began the breakdown process. Periodically, by batches, tanks were emptied into the wetland plant lagoons where canna, hyacinth, and a dozen other plants purified the water as they grew. That particular day Mark unloaded part of the wetland lagoon to make room for new wastewater. Before starting the pumps, he checked with Jane to see if the agriculture irrigation tanks were ready to receive the treated wastewater.

Sally continued her round of vegetable harvests, thinning beets from one of our new stairwell planters and picking tomatoes from plants in tubs on the bases of the space frame pillars. Linda had been processing wheat from our last grain harvests for the past couple of weeks in the basement. The noise of the thresher prevented her from hearing the radio, so she told her 'buddy', Gaie, to cover any radio calls for her. It was a big world, our three-acre Biosphere, with deep waters, cliffs and hills, as well as a basement filled with mechanical and electrical equipment, so staying in radio contact was important. Having a buddy system (divers use a similar system) helped everyone keep in touch.

That day, we planned to experiment with some new varieties of lablab beans that were given to us from a research center in India. Roy was collecting beans from the first plot in which we'd tried them, and pruned them to encourage more flowering. Jane and Taber also pruned the sweet potato plants to stimulate the growth of tubers, having collected the fifty pounds of high-protein fodder we needed daily for the goats. The weeds they removed went to the compost machine, and the sweet potato greens went to the animal bay as extra fodder.

Gaie usually spent her first morning hour tending the orchard, harvesting papayas and figs, pruning citrus and guava trees, and performing checks on the marine systems before joining the agriculture crew. This included looking over the mechanical systems, as well as data such as temperature, salinity, nutrient, and pH levels. She would then take our little boat 'out to sea' to skim the leaves from the savannah cliff face off the surface of the water, and then clean the protein skimmers, another system that removed excess nutrients from our ocean's water by bubbling air through pipes.

By a quarter to ten, Laser and Taber were at work on technical maintenance. They cleaned the filters on the basement air handlers that controlled climate in the savannah and installed new parts to the system that collected condensed water from the space frame glass over the wilderness areas. Gaie, Jane, Mark, and Sally continued with the peanut harvest, and while Mark and Gaie pitchforked the peanut plants into piles, Sally and Jane stripped the roots and piled the greens into separate buckets to be taken to the drying ovens at the end of the crew. These peanut greens would be used for animal fodder after weighing and drying,

At that point, Linda would have finished threshing and put the wheat grain into the drying oven. After drying, buckets of wheat grain go to the seed cleaning machine for the final separation of remaining leaves. Roy spent this second hour in the medical laboratory on the mezzanine floor of the habitat, working on a paper detailing the oxygen depletion studies he and Taber were carrying out. As work crew ended, the rest of us grabbed a food bucket or sack to carry to the elevator which went up from the orchard portion of the IAB to the dining room.

Although we had no physical contact with anyone except the other seven people living inside, parts of our lives were very public. Often groups of visitors would gather to watch us through the glass windows that separated our two worlds. Being caught in undignified positions no longer bothered us. Being without self-consciousness allowed us to work barefoot in shorts, even in the most amusing circumstances, such as splashing through the rice paddies splattered with mud while diving for the tilapia that grew there. We couldn't hear what people said through the glass unless our ears were pressed against it and they were shouting. Sometimes people held up signs wishing us good luck, and it gave us a boost to see their smiles and thumbs-up gestures. At times when someone wanted total privacy, it wasn't difficult to find seclusion deep within the vegetation of the rainforest, far from the side glass, or to do chores early in the morning or in the evening when visitors weren't around.

At 10:45, a break was announced over the radio. Mint tea and roasted peanuts were set out in the plaza. For fifteen minutes, we relaxed on the carpet and cushioned benches, or on the first steps of the tower stairway, then dispersed to go on to the next of our biospherian tasks.

This was Gaie's day to put on her wetsuit and scuba gear to 'garden' the coral reef: a weekly dive that included examining the overall health of the reef and fish populations, cleaning the underwater viewing windows, and removing excess algae (similar to weeding). An on-site security guard would regularly watch her through the window, acting as her diving buddy, as we were too few to have her accompanied each time. If anything went wrong, the guard would call Laser to join Gaie underwater.

While Gaie checked the coral reef, Sally moved the day's vegetables and a week's supply of staples (grains, tubers, and fruit) into the back kitchen to be weighed, logged into notebooks, and stored in refrigerators, grain bins, or the freezer. Then Gaie and Jane returned to the agriculture area to examine the crops for insects and disease. Sally had recently released two new types of mite predators, so she also collected a few sweet potato leaves to check on how they were doing. A microscope sat on a counter in the laundry room off the plaza for these examinations. Jane sprayed the rice paddies and grain fields with B.T. (*Bacillus thurigensis*), a useful bacterium which parasitizes looper worms, a harmful caterpillar, from a five-gallon backpack sprayer. This method had successfully kept the loopers under control.

By then Linda and Taber, into their second hour of wilderness operations, moved to the upper savannah to prune back the passion vines and cowpeas that were giving too much shade to the African acacia trees along the savannah stream. In the sand dune area of the desert, Mark took soil cores to measure soil moisture. On the dune sat a squat Plexiglas box nicknamed 'R2D2' (after the Star Wars robot), a device which continuously measured the CO_2 (carbon dioxide) coming out of the soil. We moved the machine from biome to biome, to monitor how CO_2 efflux changed with the seasons. The desert had its last rain a few weeks prior, and the soil had rapidly dried out, with CO_2 emissions having dropped considerably. Mark took samples from the first and second foot of soil and weighed them wet before drying them in the agricultural drying ovens. In seven days, he would have to weigh them again to get accurate soil moisture readings.

Laser continued his round of preventive maintenance, which included cleaning filters in the water systems. He called the Energy Center to check on the status of repair of their backup generator and on our supply of chilled water. The higher the outside temperature, the colder the water had to be in order to cool down the Biosphere.

We had already cut the dry grasses in the dormant savannah in anticipation of the first rainfall to bring it out of dormancy. The timing of these seasonal climate changes were determined by the SBV research division which then required coordination between the Biosphere 2 crew and personnel in Mission Control to program the air handlers. Between our morning break and lunch, Roy re-calibrated sensors for atmospheric temperature and humidity.

At 11:30 Sally started a phone call with school children in Ohio. Seated in their school library, some sixty eighth graders listened on speaker as Sally talked about Biosphere 2 and then answered their questions. Hundreds of schools around the country have used educational materials from Biosphere 2 to learn about biomes and global ecology. Many grade-school students have designed and even constructed their own model Biospheres complete with plants and even insects. All eight of us frequently connected with schools in Arizona and around the country—one of the most enjoyable of our "jobs" while inside. When Biosphere 2 was first planned, it was not expected that it would be more than a quiet research endeavor in the somewhat remote, and seasonally quite hot, Sonoran Desert of southern Arizona. We expected to recoup the investment by building other biospheric laboratories in world cities which would double as eco-tourism destinations. But when Biosphere 2 struck a nerve and excited people around the world, people began coming— by the dozens, hundreds, then thousands, even before we had created a formal visitor's program. Consequently, we realized that biospherics education had to be an important part of Biosphere 2's program; and reaching and changing minds, especially young minds, was a major payback for the epic creation of the facility.

By noon, Gaie showered, changed, and returned to the kitchen to make final preparations for lunch. Like most of the Biosphere 2 cooks, she'd done much of the work the previous afternoon while fixing dinner. With the vegetables cut and the potatoes already baked, she only needed to finish the salad. She made a dressing (bananas blended with water and chopped herbs), popped her baked potato casserole into the oven to reheat, and salted the soup of beans, vegetables, and chicken broth which had been slowly cooking in a crock pot since the night before. She then sautéed Swiss chard and beet greens in the wok and mixed up another cold salad of sliced beets and papaya.

Jane and Sally were giving the goats and chickens their midday feed. The buck, Buffalo Bill, received a diet lower in protein than the milk goats and had to be locked into a side pen during each of the three daily feedings. The two kids normally kept Buffalo Bill company, but since they were now a week into being weaned, they also were fed a different diet in a separate pen. Buffalo Bill was a veritable Houdini at opening pen doors, so before leaving, Jane routinely double-locked them and tied them with a loop of baling wire.

Lunch time was 12:30. A couple of latecomers had let us know via radio that they'd be along soon, and we prepared plates for them. All our meals were either served on individual plates or put out in servers where it's easy to see each one-eighth portion. *Everything* was eaten, a tribute both to the care with which the cooks prepared the meals, and to the appetite we brought to each meal.

Over lunch, we'd share more news. We were all especially interested in animal sightings in the wilderness, news from TV, radio, or email, or what comes over the grapevine from friends or staff with whom we had recently been in touch with by phone. Jane brought up the coming arts festival we had planned and who on the outside would be sending in their music, paintings, or poems over the video system. Laser enthusiastically described his encounter with a baby galago in the orchard the previous night; Linda said it was a baby newly born to Topaz. One of our buddies in Mission Control took on the weekly task of renting the latest video releases for us. Laser announced that *Lorenzo's Oil* and *Malcolm X* had just been piped in and we all let out a hearty cheer. 'Piping in' a movie means that Laser would place a blank tape into the VCR to record what had been transmitted through the video link with Mission Control.

For an hour and a half after lunch, we usually took an informal siesta. After the initial months of adjusting to the new diet, the heavy physical work, and the lower oxygen supply, not many of us actually used the time to take a nap, but instead relaxed in our rooms reading or phoning friends. It became a welcome break in the workday. Everyone in Mission Control knew our schedules, so there were no radio calls during that time, unless there was something urgent to attend to. This day Mark's family came to visit him at the special window we use to meet with outsiders. His mother, who after closure, made the big move from Brooklyn to Tucson, his brother, and his sister-in-law, who lived just a few miles away, chatted with him at the window through a speaker phone.

Around 2:30, Sally would place the next allotment of food in large plastic tubs in the refrigerator. Taber, the next cook on the rotation, began his cooking duties by checking the blackboard in the back kitchen where Sally noted the available quantities of some of the staples. These changed from week to week depending on our harvests and on Sally's calculations (assisted by computer) of our nutritional needs. Taber's beans had already been soaked overnight, and since the morning they had been slowly cooking in a crock pot. Our foods included many whole foods, so there was more washing, peeling, and soaking than normally required in a typical American meal preparation.

Jane and Gaie put on loose-fitting work clothes for the messy, sweaty job of cleaning two rows of algae scrubbers. Gaie cleaned the Plexiglas containers and wave-buckets, while Jane scraped the old algae off the screens. The harvesting of the algae off the screens was labor-intensive, so the biospherians would alternate the job, each doing one to two hours every two weeks. Before they completed the job, Jane and Gaie placed the scraped-off algae in racks in the room's drying ovens.

Now R2D2 had to be moved to its new location in the lower savannah. First Linda disconnected the long cable which carried its measurements to one of the large computer stations in the basement. Then Mark and Taber coiled the cable and extension cord and gently moved the 'portable' but awkward sixty-pound apparatus over the surprisingly rugged and varied Biosphere 2 terrain. They re-ran the cable and extension cords to the nearest side air vent that drops down to the basement and Linda later rewired the cable to the closest computer cabinet and electrical outlet. There were computers along with thousands of sensors all over the Biosphere: parts of the "Nerve System" that collected data so it would be available to us and Mission Control.

In each biome, a circular plastic ring had been fixed in the ground for R2D2 to sit on. A greased gasket ensured an airtight fit. The instrument had two sensors to measure both the carbon dioxide in the atmosphere and the carbon dioxide diffusing out from the soil. Before R2D2 was constructed by Mission Control staff at the end of our first year, Linda and Taber used to take the measurements manually, taking a syringe of air every six hours over twenty-four-hour periods and then running the air samples on a gas chromatograph in the Biosphere 2 laboratory. We'd occasionally use the manual method to check the accuracy of R2D2 and to take samples in other biomes if R2D2 was occupied elsewhere.

The weekly Mission Control meeting via video started at 3:00 PM. Sally and Laser joined Norberto Alvarez-Romo, head of Mission Control, Bill Dempster, in charge of engineering systems, and Bernd Zabel, operations manager for the Biospheric Research and Development Center (BRDC) to coordinate activities between the two worlds and review technical systems. Sally and Laser sat at the V-shaped table in the command room which was outfitted with both the video link and a document reader. They talked over the operational and research activities for the week and a few problems that needed Mission Control's attention. Mission Control noted that vegetation was pressing against the space frame glass at several points and needed pruning, and that the glass also needed cleaning at several places.

They made sure that we knew about all upcoming visits by collaborating scientists, guest lecturers, or VIPs in the environmental world.

During the afternoon, we usually worked on our weekly and monthly reports or contacted people with whom we had been doing research projects. Sally finished her weekly report on food production, diet, and nutrition; Jane reported on animal fodder production; and Laser detailed the maintenance program. There were also reviews and projections for the terrestrial wilderness systems (Linda), the marine wilderness systems (Gaie), and the medical and health systems (Roy).

By 3:30, the Biosphere would be bustling. Laser was working on technical design problems and the future upgrades during the transition (the period after the completion of the two-year closure when system improvements, detailed research with outside scientists, and helping and training the second crew would occur) with Ernst Thal-Larsen and Larry Pomatto, Director of Technical Systems, via video. Sally was on the phone with Jim Litsinger, the project's integrated pest management consultant, and Dr. Michael Stanghellini, a pathologist at the University of Arizona who had recently examined roots of our rice, wheat, and sorghum crops to see if nematodes or water-borne bacteria were responsible for the poor yields. They discussed a test which had been proposed by consultants at the University of Michigan to see if our rice seedlings were suffering from a more subtle nutrient deficiency.

Mark was on the phone, too, discussing an upcoming meeting, The Case for Mars Conference, with its organizers in Boulder, Colorado. Using PictureTel technology, we were able to present a 'paper' detailing our findings and also participate in a direct discussion with the conference members. After he hung up, he made data entries from the leaf litter and decomposition studies underway in all the biomes.

Within the hour, Gaie was scheduling the first three months of the anticipated transition phase (the period between Mission One and Mission Two from September 26, 1993 to February 26, 1994) that would provide the opportunity for research, review and repairs inside the laboratory. During that period of time, Biosphere 2 would continue to operate with closure protocol. Work teams were prioritized and planned according to the breath of an eight-person crew living inside. Thus, this task needed to include all sixty research programs underway, as well as technical system repairs or upgrades, and training the second crew that would operate the laboratory for the coming year. In order to set the research plan, she contacted a number of the consultants and collaborating scientists by phone and email to arrange a meeting within the Biosphere to make measurements and observations for the completion of research projects.

Taber spent the afternoon doing routine maintenance on the automatic sniffer system that analyzed levels of eight critical gases in the atmosphere.

It may sound as if everyone was so involved in activity that there was little room for emotional interaction. But there was—even if conducting our private relationships proved a bit tricky. We agreed not to invade in one another's personal lives by addressing who was or wasn't involved in love affairs, suffice it to say that there were some people in Biosphere 2 with such relationships, and others without them. For those without, how was it possible to continue a relationship with someone on the outside? With ingenuity. Private meetings at the windows were arranged. While the official biospherian handshake was two hands matching each other (even though separated by the thin three-eighths-inch glass), pairs of lip-prints had been spotted on each side of the meeting window as well. Sometimes they took a while to fade, lingering like the blush of flowers preserved in a diary.

At 4:30, Linda and Mark met in the west arroyo area of the desert to measure plants. This was part of a biomass resurvey they completed at the end of each rainy season, to measure how the communities of plants were changing as the desert developed. At the same moment, Sally was discussing the results of recent trials of recipes from her Biosphere 2 cookbook, *Eating In*, with one of the 'food testers.' Gaie had just completed her daily review of the Biosphere 2 management with John Allen, Director of Research and Development, and was now on the phone with Dr. Jack Corliss, Director of Research for SBV, about the overall research program which included wide areas of investigations in global modeling, biogeochemical cycles, biomes and ecosystems, systematics, human physiology and nutrition, and engineering.

Meanwhile, a last burst of animal husbandry was taking place: by 5:30, Jane was milking goats, Taber came down from the kitchen to put the buck and kids into their separate pens to feed them, and Sally was taking care of the chickens. After completing the biomass measurements in the desert, Linda connected the cables for R2D2 and once more checked in with Mission Control regarding CO_2 emissions. Mark went to take more soil moisture samples.

By now the official workday was over, but, as usual, a few matters needed attention. Laser was trouble-shooting the electronic sensing devices of the fire alarm system which had been giving false alarms due to high humidity. Gaie and Mark reviewed a press release written by the public affairs office. Sally returned a call from a Tucson journalist requesting her reaction to the National Academy of Sciences warning of the danger of pesticide levels in fruits and vegetables, especially to children. She used the opportunity to discuss the relevance of our agricultural system, which didn't rely on chemical pesticides or harmful chemicals of any kind and was nevertheless extremely productive. The journalist also wanted to come out for a photograph of Sally with some of her new beneficial insects (which eat crop pests), so she arranged a time for the following morning.

Roy was cook's helper for the day and began helping Taber by washing pots, making tea, and setting the table in the frequently frantic last minutes before meals were served. Gaie, Mark and Sally, hungrily anticipating dinner, took the opportunity to work on this book, which they'd been writing since the early days of the experiment.

Taber and Roy served dinner at 6:45 PM. Taber made one of his characteristically beautiful meals, featuring a flour shell into which he placed delicious chili with lablab beans, chives, taro, and green banana. There were sliced beets, a mixed vegetable dish, baked sweet potatoes with a banana sauce, and tossed salad. Dessert was sweet potato pie topped with slices of fresh fig.

Less than an hour later, Taber and Roy were doing the kitchen cleanup, which included sweeping and mopping the floor. They cooked the day's food scraps in a large pot on the electric stove, and these leftovers were now ready to go down to feed the goats. Taber started on the next day's breakfast. He began slow cooking the porridge overnight in a crock pot, and preparing rolls and sweet potato patties on baking sheets in the refrigerator along with other foods for tomorrow's lunch.

Sally and Laser worked together in the command room summarizing the day's events in the captain's log and sent it via the network to Mission Control and SBV management. At 8:15, Mark started his night rounds—a twenty-to-thirty-minute tour through the agriculture, rainforest, savannah, and desert, noting the manual thermometer readings in each area. He also turned off whatever lights that had been left on, and then checked the algae scrubber room, the recirculating fans, and the wave machine in the ocean. He also glanced at the alarm screen in the savannah tunnel for any temperature red alerts, and then checked another alarm board nearby for any yellow lights indicating technical malfunctions. He then copied his readings in a logbook which was kept near the alarm screen monitor.

Linda settled in with her computer to log onto the WELL, a communications network run by the Whole Earth Catalogue. She checked a general bulletin board for useful information and examined WELL's array of electronic conferences. Linda frequently contributed to conferences on Biosphere 2, as well as many others on politics, the environment, and cutting-edge technologies. Roy was on the phone with friends in Los Angeles. Taber and Jane were watching TV, and Gaie was reading.

Our apartments were our castles. When we wanted to be alone, that is where we'd retreat. Each biospherian had a two-story apartment with a downstairs area that includes desk, bookshelves, TV/VCR and radio/CD/tape player, sofa and easy chair, and clothes closet. Up a circular staircase is the loft bedroom with a clothes bureau. There were ten apartments in Biosphere 2, the extra two were allocated for visiting scientists who may use them in the future.

Before closure, the crew had the opportunity to decorate their 'pads' with whatever they wanted. So, we had our favorite paintings, souvenirs from travels around the world, books, and music collections. We also chose rooms with views that we liked—though all the views were out of this world. Some looked out over the agriculture area and south to the Santa Catalina Mountains, others looked out over the ridge to the west and were good for sunset-watchers. Some had a view of the area in front of the habitat, where gleaming, white tents sprung up during special events at Biosphere 2.

Laser transformed his apartment into a video studio with gear for making documentary films. A piece of blue plastic in his window mystified visitors until he explained that he was protecting his equipment from harsh sunlight. Linda decked out her apartment with beautiful artifacts from native cultures around the world. Roy's apartment was decorated with art pieces from artist friends. Mark's was dominated by his huge library of books, Australian Aboriginal carvings, and colorful paintings.

Most of the crew kept a personal record of their experiences, and before turning in at about 9:15, Mark made his daily entry in his computer. Others kept long-hand notebooks. By 10:00 PM, the hallways of the habitat were dark, although light seeped out from under the doors of a few apartments.

For all eight of us, another day of the 731 days we would spend in Biosphere 2 had come to a close. As we drifted into sleep, our world continued to hum around, above, and beneath us. Almost all of the crew have remarked upon the special intensity of dreams inside Biosphere 2, but no one noted any particularly unusual dream that night. A few of us may have noticed the sliver of the new moon rising outside, refracting through the beautiful geometric glass sky of our miniature world.

Mark Van Thillo calibrates a humidity sensor in the savannah.

CHAPTER 3

MONITORING THE ENVIRONMENT

"Biosphere 2 provides, for the first time, the possibility of conducting controlled, large-scale ecosystem ecology experiments. Modern Physics emerged when Galileo conducted experiments that yielded numerical data. Biosphere 2 provides a setting for the same type of transformation in ecology as occurred in physics."

– **Dr. Harold J. Morowitz**, Robinson Professor of Biology and Natural Philosophy, George Mason University

OUR OWN EPA

IF THE CITY SEEMS SMOGGY, maybe you can drive out of it to the country. If it's smoky near a factory, you don't go near it unless you live or work nearby. Oil spills and nuclear plant leaks—well, they're nasty, but maybe they're far away and there are other things to worry about. But in Biosphere 2, this way of thinking would never do. In a small living system, it is vividly clear that everything we did had immediate and potentially devastating effects on our surroundings. So, we policed the environment, ever alert for anything that could pose a health threat to ourselves or to any of the other life forms. That meant paying especially close attention to the quality of air and water, the first of the critical life support elements to reveal the onset of pollution.

Part of the pre-closure preparations involved the formulation of 'mission rules.' These resemble the rules that might be outlined by an environmental protection organization. Among the guidelines were air and water temperature limits, the allowable concentrations of potentially toxic molecules or heavy metals in the drinking water or air, and the management of nutrient cycling. There were many unknowns, and many more partly knowns, in this process. After all, the guidelines can only be set by measuring the natural world in minute detail, the means of which have only just begun to be developed during the past few decades. In the case of specific air and water contaminants, the SBV research adopted the standard warning levels used by OSHA for industry, by EPA for the environment, and by NASA for astronauts. One of the long-term purposes of the Biosphere 2 experiment was to help develop better environmental guidelines for use everywhere.

Management of our closed ecological system required very precise analyses of our environment. We depended on getting those measurements quickly enough to make the right decisions by a combination of electronic sensors, automated analytical devices, and manual laboratory procedures. Biosphere 2 was equipped with a state-of-the-art analytics laboratory capable of making the precise measurements of small concentrations of compounds in either air or water, designed to run without relying on outside support, consumables, or chemicals. All other existing laboratory systems used large quantities of toxic chemicals that we just couldn't release safely into our living system. So, working with experts, SBV innovated by creating a lab that used no toxic chemicals and recycled any by-products in a constructed wetland waste recycling system similar to the one that handled the wastewater from the human habitat. In this way, we were able to avoid the paradox of most environmental laboratories that, even as they helped identify toxic pollution problems, they created pollution themselves.

Before the start of closure, project management created a Scientific Advisory Committee (SAC) of distinguished scientists from a variety of disciplines to provide scientific consultation for the Biosphere 2 experiment. The SAC provided periodic guidance on research as well as outreach to other consultants as the research program developed. Preparing for their several meetings held at Biosphere 2 consumed a lot of time, especially for Gaie, as our Director of Research inside, but also for Roy and Mark who helped Gaie organize the meetings and report on the 60 research projects underway. Keeping necessary operations going during these 2 and 3-day meetings was challenging for all the crew since everyone was involved in research work.

After the first ten months of closure, the Scientific Advisory Committee issued a report which lauded the project as "an act of vision and courage" with the potential to advance our understanding of the biosphere such as biogeochemical cycles, ecological patterns and processes, sustainable agriculture, and atmospheric dynamics. They also made recommendations for ways to lessen biospherian work and enhance the research program. For example, the SAC recommended that we export research samples through our airlock for analysis in SBV's outside lab and other independent laboratories. This way we could increase the amount of data we were getting from the experiment and also decrease the time we spent doing the analyses inside. Thus, after fourteen months of closure, SBV made the decision to export much of our laboratory analysis equipment and set it up in our Biospheric Research and Development facilities outside Biosphere 2. So even though the merit of the new laboratory system was proven, in the end it proved more efficient, and biospherian labor-saving, to have this superb equipment operated full-time on the outside. Additionally, the SAC suggested that an overall Director of Research be appointed so as to expand the research program on the outside and augment all the research programs after the completion of the first two-year experiment. Thus, in the summer of 1992, Dr. Jack Corliss, a modeler and pioneer in discovering the deep ocean hyperthermal vents who had been a senior researcher at the NASA Goddard Space Flight Center, joined the team as Biosphere 2's Research Director.

The wonders of modern analytical technology allowed most air and water compounds to be detected even in trace amounts. Some can even be detected when they are as low as a few parts per billion (ppb) or even parts per trillion (ppt)! Combining these analytic techniques with the fact that Biosphere 2 was a closed system, enabled us to track compounds and detect increases of toxic compounds long before they approached threatening levels. These advanced systems also enabled us to do some fascinating detective work.

A few months after closure we experienced a case in point. Trace gases, which commonly come from the glue used to seal PVC pipes, were persisting in our atmosphere. Yet, from our earlier experiments, we knew that these gases should have been gradually cleaned up by natural ecological processes. This meant that there was something inside the structure that was still releasing the gases. We immediately searched Biosphere 2 for any containers of glue, primer, or sealant that may have been forgotten by the construction crews who put the pipes together. To our amazement, we found four separate containers of PVC glues and solvents, in remote corners, which were slowly leaking fumes into the air. Once these cans were sealed off in containers and removed to the west air lock, the levels of the trace gases we were tracking in the atmosphere finally declined.

Because there were so few sources for toxins in our atmosphere (we had no gasoline engines, cigarette smoke, paints, synthetic carpets and materials, or even perfumes), it was primarily welding and technical repair operations involving PVC solvents or glues that caused small spikes of trace gases to appear. During the first year, the SBV research and analytical staff debated whether the atmosphere and our living ecological systems, which ultimately metabolize the gases, could support necessary repairs to some of our mechanical systems. The discussions pushed us to devise strategies to minimize their impacts on our atmosphere.

In fact, we had to think about these things all the time, about everything we did. The consequences of seemingly innocuous actions like opening a can of PVC glue are immediate and real. There are no 'anonymous' actions. Even if someone didn't tell us that they did some midnight painting or gluing, we'd know by observing which trace gases had risen in our air since the last analysis.

SNIFFING AND SIPPING

In order to measure some of the gases contained in our atmosphere, SBV set up an elaborate device called the 'sniffer system'. This was an automated system located in our analytical lab in the habitat with sensing stations located throughout the Biosphere. These continuously monitored the gases that could cause health concerns as well as those which we thought would show significant changes during the experiment: CO_2 (carbon dioxide), O_2 (oxygen), NO_2 (nitrous oxide), NO_X (other nitrous gases), SO_2 (sulfur dioxide), CH_4 (methane), O_3 (ozone), and H_2S (hydrogen sulfide). Each gas is detected by an instrument that 'sniffs' the air every fifteen minutes. We and the project's research staff were able to access current values from the data base or generate graphs to show trends over time. To measure other trace gases, we collected samples of the atmosphere every two weeks and analyzed them with the equipment in our lab or sent out to collaborating labs.

To lessen any potential problems to human health from the gases, the Biosphere 2 design team chose to use natural building materials such as wool and wood. We also tried to avoid materials known to have nasty trace gas emissions. Although we occasionally had to use glues, sealants, or grease to repair piping or maintain pumps, we made an effort to keep their use to a minimum. Before this experiment, however, no one could fully predict the magnitude of the problems we would face. This was the first time such a complex system of technology and ecology had been sealed in a nearly airtight facility. For example, we wondered what methane would do because our world included a mangrove marsh biome, rice paddies, compost heaps, constructed wetlands, and even our own digestive systems—all methane emitters. Our microbial ecologists predicted that the methane-eating bacteria in the soils would increase to keep the methane in balance. And indeed, after an initial period of increase, levels of methane stopped rising and remained safely low and steady throughout our two-year closure.

Although we had over 100 trace gases in our atmosphere, the only one that had slowly but steadily increased since closure was nitrous oxide, although it was still at safe levels. This was the subject of much humor among us, as nitrous oxide is also known as 'laughing gas'. But though we have about thirty times as much nitrous oxide as Earth's environment, it was still measured in tens of parts per million. To get the full dentist's chair effect, it'd have to be well over half of our air. It's interesting that nitrous oxide was our only problem trace gas. Nitrous oxide is only broken down by UV rays high in

Earth's stratosphere, not by microbes like most trace gases. Biosphere 2 lacked a stratosphere, of course, even though our architecture featured very high ceilings.

But what about the smells? Even minute amounts of a gas created a stink or a fragrance. Fortunately, we had the opportunity to verify that we had not just become accustomed to our air, thinking it smelled fresh when, in fact, it really didn't. During our import and export exchanges of research materials, those on the outside waited eagerly to come into the airlock chamber to take deep breaths and catch a whiff of our atmosphere. They reported that the smells were delightful, similar to the deep organic smells of rich farmland or a rainforest.

The variety of natural smells was one of the many delicious surprises we encountered when we conducted our first Test Module experiments. (The Biosphere 2 Test Module was a 17,000 cubic foot [480 cubic meter] 'prototype' where we did some of our first experiments before building the much larger facility). People who lived inside the structure for days or weeks always describe the rich natural smells as a freshly cut field of hay or a rich, earthy aroma. Even more intriguing, all the biospherian trainees and SBV staff, who spent a minimum of twenty-four hours to a maximum of five days inside the module, commented about how smells differed depending on the time of day.

A variety of aromas was deliberately included in the design of Biosphere 2. All the biomes had different smells as well as different moods, light, and humidity. Many of the plants in the desert were chosen specifically for their pungent aromas. These aromas may either be part of the plants' chemical defense systems to discourage grazers from munching precious leaves or part of its attracting system to lure pollinators. For us, they made a trip to the desert a sensory feast. It was a delight to brush against the plants and release the aromas. In the smell-conscious environment of the Biosphere, other crew members could always guess where you'd been when you came back to the dinner table.

For monitoring critical water parameters, we designed an automated 'sipper system' similar to our sniffer system for air. It was located in the technical basement below the savannah where it received water sequentially from the ocean and marsh and analyzed the concentration of phosphates, nitrates, and nitrites. These nutrient measurements were critical for the management of coral reef and marsh health.

Coral reefs are often referred to as the 'rainforests of the ocean' because of their dazzling biodiversity, vibrant colors of coral, and aquatic life. But curiously enough, coral reefs thrive in extremely nutrient-poor waters and are quite sensitive to changes in nutrient content. An interesting example of this occurred pre-closure during the summer monsoons in 1990 when the ocean had been stocked with Thalassia lagoon grass from Florida and reef rock turf from the Caribbean. The monsoon season in the Sonoran Desert brings heavy rainstorms. Since this section of the Biosphere had not yet been sealed overhead, the rains often added between 6,000 and 12,000 gallons of unwanted water into the fledgling ocean at a time. Invariably the ocean water turned a pea-green color the next day, an indication of an overabundance of plankton. We found that the rainwater was quite laden with nutrients, with significantly higher nitrogen and phosphorous concentrations than are found in reef water! Even though the city of Tucson was about thirty miles away, nutrient-laden smog contaminated the rainfall.

The relative acidity or alkalinity of the ocean is described by a pH number. A pH of seven being neutral, lower than 7 being acidic, and higher than 7 being alkaline. Initially, the ocean pH was targeted to maintain levels within the range generally found throughout the world's reefs, from pH 8.2 to 8.4. But the Biosphere 2 ocean could not maintain these levels because of the higher and frequently changing levels of carbon dioxide in our atmosphere. The ocean water absorbed carbon dioxide, which lowered the pH of the ocean. So over time, we had to lower the alarm thresholds, the lowest was 7.6, since we were often operating at pH levels lower than those in natural conditions. We were relieved to see that the coral reef community tolerated the changes. In fact, much of what we learned centered on redefining ecological tolerances as we discovered how the biomes reacted to conditions not naturally found in the environment. By recreating complex biomes isolated in Biosphere 2 and exposing them to new environmental conditions, we were experimenting with ecology. Ecology had mostly been an observational science; Biosphere 2 was the first facility to make ecology at this scale an experimental science. *That,* to us, was the really exciting potential being realized.

Our drinking water was condensed out of the atmosphere, either collected from the water that condensed on the space frame glass or caught in trays in the air handling units. Depending on the season, we had 4,000 to 8,000 gallons of pure water each day. This was possible because water moved quite rapidly through our atmosphere by means of the transpiration and evaporation cycle. Transpiration is a process by which plants pass enormous quantities of water through their leaves. For example, a corn plant will transpire an estimated 100 gallons of water for each gallon that it actually uses for growth and fruiting. In addition, the large surface area of the ocean and marsh allowed a considerable amount of water to evaporate, which also increased our atmosphere's humidity. On days when the humidity was very low, we estimate that approximately 2,000 of the ocean and marsh's 972,000 gallons of water evaporated. Over the course of a year, almost all of our ocean would evaporate and be replaced by other Biosphere 2 water! During this process most salts and other trace compounds were left behind, so the condensate water we collected was quite pure and suitable for drinking.

Once the condensate was collected, we would pass it through an ultraviolet/ hydrogen peroxide system to kill bacteria. We chose this kind of sterilizer, rather than one using chemical additives, because our drinking water would eventually wind up in our wastewater wetland lagoons, the irrigation supply for the farm, as well as all our wilderness biomes. Each of these biomes had their own water quality, thus the process of preparing rainfall tanks in each area required the proper mixing of grey water with condensate water. Even small amounts of chemical purifiers, such as iodine or silver, would build up to dangerous levels over time. On the scale of a century, the planned lifetime for Biosphere 2, this would become an unacceptable option for it would eventually prove disastrous. Our internal EPA rules took this longer view very seriously, as we didn't want to pass on any environmental timebombs to future crews.

All of these safeguards put together had given us a safe, continuously viable system. We conducted a full analysis of our drinking water twice a month, while the waters passing through our soils, and the wilderness condensate, streams, and rice paddies were tested once a month.

OVERCOMING THE SICK-BUILDING SYNDROME

Air is complex. Although it mostly consists of nitrogen and oxygen, the small concentrations of other gases it contains are very significant for the living systems that breathe it. The primary gas components of air are nitrogen (78.08%), oxygen (20.9%), argon (0.93%), and carbon dioxide (0.035%), but there are significant though small amounts of what are called trace gases, important because some may become toxic if their concentration rises even slightly.

You don't have to investigate the Biosphere 2 experiment too deeply before it becomes obvious that maintaining a healthy atmosphere was our most important task. Even with our high roofs, there's a relatively small amount of air in our world compared to the mass of soils, plants, and man-made materials and equipment. Our experiments in the Test Module showed that even rigorously evaluated materials still produced some outgassing into the atmosphere. This outgassing could produce a bewildering mix of chemicals. Even in spacecraft (such as Skylab or the Space Shuttle) where materials were chosen to minimize outgassing, hundreds of trace compounds had been detected in the cabin air, just as we found in the Test Module.

Trace gas build-up has always been one of the most serious potential problems in small life support systems. Before closure, Mark spoke with a space scientist who told him that when the space shuttle's doors were first opened at Edwards Air Force Base after a week in space, a member of the ground team threw up because the smell inside the spacecraft was overpowering. The obnoxious smell came from the notorious zero-gravity toilet, which then combined with other odors accumulating in the tightly sealed small cabin. In the Russian Bios-3 experiments in Siberia, even the plants once stopped growing due to the toxic buildup from algal tanks.

These types of problems aren't confined to spacecraft—many ordinary homes and office buildings suffer from the 'sick-building syndrome'. People quickly get used to bad odors—so much so that we often have to be informed of a smell by someone with a 'fresh nose' who has just entered the environment. It's not just that the smell is unpleasant, it can also be unhealthy. In many modern houses and buildings, which are tightly sealed for energy efficiency, the indoor air can be far more polluted than the air in the streets. The interrelationships of plants and soil microbes are beneficial in that they are uniquely able to clean up these trace gases. More widespread use of indoor landscape planting areas in offices as well as of potted house plants would go a long way towards solving the sick-building problem.

To ensure that we would not be plagued with these problems in Biosphere 2, we developed and patented a system based on an engineering technique first developed in the early 1900s. A discovery made in Europe showed that pumping the nasty odors from food-processing factories through a nearby field of soil purified the air. They called this a 'soil-bed reactor'. SBV turned this discovery into an ecological apparatus that purified the air of any toxic elements through the form of a small commercial unit, called an Airtron™ that cleaned the air using the combined activity of houseplants and their soils. Soil biofiltration of air pollutants was a major incentive for the Biosphere 2 designers to make our farm a soil-based rather than hydroponic system. This soil-based farming system, in addition to allowing us to recycle organic wastes through composting techniques, also allowed us to control toxic gases in the Biosphere 2 atmosphere.

At first, a small portion of the soils in all the biomes were designed as soil bed reactor (soil biofiltration) back-ups. But finally, to be able to deal with any eventuality, the entire agricultural area was constructed with an Airtron™ system to allow air to be pumped through all the soil. If necessary, we could force the entire Biosphere 2 atmosphere through the farm soil in about a day. Part of the beauty of the system is that whatever trace gases exist in the air serve as food for microbes and these microbial populations would increase until they have brought down the concentration of trace gases. The increased air flow can even help minimize pockets of waterlogged or anaerobic soil. It's a marvelous technology born of billions of years of microbial evolution!

Thankfully the natural systems in Biosphere 2 had functioned so well that we hadn't any need to use the soil-bed reactor inside Biosphere 2. For us this was a welcome victory because we were also realizing through our studies with R2D2, that activating the soil bed reactor would also result in flushing CO_2 into the atmosphere because soils also respire. Thus, the management of trace gases in Biosphere 2 was balanced and we only activated this system for test and research purposes.

THE TIGER OF BIOSPHERE 2

Because of the much smaller ratio of atmosphere to soil in Biosphere 2 compared to Biosphere 1 (Earth), the design had to take into account that soil respiration could, at least in the first few years, increase CO_2 levels to daily averages over 4,000 parts per million during the winter months. In addition, the smaller atmosphere meant far greater daily changes (over 600 parts per million) of CO_2 than the one or two parts per million daily fluctuations observed in Earth's atmosphere. So, we paid very close attention to carbon dioxide.

On a normal day of sunshine, CO_2 could drop by as much as 600 to 800 parts per million and then rise as much during the night. By comparison, global CO_2 only varies by ten to fifteen parts per million over the course of a year. But just as the citizens of Earth are concerned about the slow and steady increase in CO_2 caused by industrial emissions and the loss of forests and other natural biomes which can absorb and store carbon dioxide (the greenhouse effect which causes climate change), the Biosphere 2 EPA also had CO_2 near the top of its list of concerns. It could cause problems either by getting so high (over 1 percent or 10,000 parts per million) that it causes animal and plant toxicity, or by dropping so low (anything much below 250 parts per million) that plants become limited in their ability to make new growth through photosynthesis. Biosphere 2 operated between around 600 and 4,800 parts per million during our two-year closure experiment as its daily average depending on the time of year. One of our key findings for both the Test Module and Biosphere 2 was that CO_2 dynamics and levels in each closed ecological system will differ from Earth' biosphere and depended on its reservoirs of air, water, amounts of soil and biomass, and local climate day lengths.

Every fifteen seconds, a new value for CO_2 from continuous sensors distributed throughout Biosphere 2 (agriculture, habitat, rainforest, savannah, and desert) appeared on the 'global monitor' computer screen. Biosphere 2 was so sensitive that simply by looking at the numbers changing on the screen we could tell whether a cloud had momentarily blocked the sunlight or a brief patch of sunshine opened up on an otherwise overcast day.

Carbon dioxide was on a roller coaster in our small world because of its sharp rise and fall from day to night and the way it quickly rose over successive days of cloudy weather. Helping the system manage the CO_2 levels was one of our major challenges. We called carbon dioxide 'the tiger of Biosphere 2', and as the saying goes, "once you catch a tiger by its tail, never, never let it go!" So, in addition to close monitoring of its levels, the research department developed a number of strategies to boost photosynthesis and minimize respiration. It became clear as we practiced these techniques, and experienced daily, how our actions made a difference, that the biospherians were becoming stewards of Biosphere 2.

Reducing respiration, which releases CO_2, also helps correct the 'greenhouse effect'. The best way we can help lower CO_2 is to maximize plant production so that the plants fully utilize the precious 'sunfall' that enters our world. As examples: plant boxes and pots were built inside during the two years to increase plant growth. Trees were pruned and vines were removed from the glass to ensure that sunlight reached the understory vegetation. This stimulates new growth that takes up CO_2 more efficiently than does old growth. Biomass was cut from fast-growing areas of the wilderness biomes such as the ginger belt perimeter of the rainforest and the tropical grasses of the savannah. The storage of the pruned vegetation kept the carbon it contained from being respired back into the atmosphere as CO_2. We also stopped composting in the agriculture during winter since this process also released CO_2. In the sunny spring season of long days, we resumed composting and returned the stored plant material to the ecosystems as mulch. Agriculture fields were tilled only to about a one-foot depth to reduce the amount of respiration. The seasonal rainfalls for both the savannah and rainforest were altered to increase biomass production during the long winter months. As well, we cooled the overall temperature of the facility Biosphere 2 to help slow its metabolism and thus reduce respiration.

To deal with the worst CO_2 rise during low light months, which we knew was going to cause challenges due to our Test Module experiments as well as nine-week-long pre-closure Biosphere 2 experiments, we built a unique recycling system which mimicked Earth's natural geological processes. The recycling system operated from time to time in the late fall and winter to extract CO_2 out of the air, and through a series of simple chemical reactions, changing it into a powdery limestone, calcium carbonate. When CO_2 was needed back in the atmosphere, this limestone would be heated in a furnace to re-release the CO_2. We anticipated that this re-release would be needed in the future when the life systems were reaching a climax ecology and thus able to re-use the stored carbon. The benefit of the system was that the recycled starting chemicals could be used all over again. This is analogous to what occurs on Earth over centuries: CO_2-storing limestone forms on the ocean floor and CO_2 is released from the Earth's interior in volcanic emissions.

The unprecedented cloudy winters during both years of Mission One proved to be as good a test as any of our ability to track and manage CO_2. In 1991 and 1992, the El Niño ocean currents caused the southwest United States to receive an unusual number of fall and winter storms. In January and February of our second year, southern Arizona suffered record floods from a series of successive winter storms. Despite all the cloud cover, our winter CO_2 levels reached around 3700 parts per million the first year and peaked at around 4800 the second year. When the sun finally returned in March of 1993 for our final six months, it was clear that Biosphere 2 managed to handle the problem, to our immense joy. The research department anticipated that the Biosphere itself would more completely self-regulate the CO_2 in future years as the young soils mature and have less organic material to produce CO_2 from soil respiration. Also, as the plants and trees grew larger, their photosynthesis would increase, helping balance total system respiration. Preliminary estimates were that the overall plant biomass (amount of living matter) more than doubled during Biosphere 2's first two years of operation.

Oxygen was also monitored closely, using two instruments in the analytical lab. One was a Teledyne oxygen analyzer, a standard piece of equipment for determining safe oxygen levels. The early readings indicated oxygen levels in the Biosphere were above twenty percent, similar to oxygen levels in Earth's atmosphere. But because this instrument was designed as a rough indicator of the rise or fall of oxygen levels, Taber was scheduled to make a routine calibration check every six months using the more sophisticated and accurate gas chromatograph.

In February 1992, five months after closure, Taber was shocked to see during the calibration check that the oxygen levels in the Biosphere were falling markedly. It became clear that the continuous analyzer for oxygen had been giving inaccurate readings since closure. This drop in oxygen levels was particularly startling because during all the Test Module experiments we never experienced a fall in oxygen levels. Nor had it been reported by other life support facilities, so it wasn't an issue anyone in the field considered critical. To learn from the unexpected, and to find out what we didn't know, was precisely why we built Biosphere 2. But the oxygen was indeed declining, and it rapidly became our number one concern. John Allen, SBV's Vice-President of Research and Development, and Bill Dempster, our systems engineer, met with Wally Broecker, geochemist at Columbia University's Lamont-Doherty Earth Observatory, and began a research program under Dr. Broecker's guidance with a young doctoral student, Jeff Severinghaus. Several other scientists specializing in geochemistry and physiology joined in. It was decided, with the consent of the biospherians, to 'ride the oxygen down' because of the extraordinary research opportunity to learn about the cycles of oxygen, and thus wait to replenish oxygen levels the moment we observed any serious health issues.

Finally, the time for a difficult decision came: On January 13, 1993, after several members of the crew began experiencing physical difficulties, including sleep apnea, a measured amount of 31,000 pounds of liquid oxygen were injected into a lung, so it could be precisely tallied, then released into Biosphere 2's atmosphere over a nineteen-day period to boost the levels of oxygen from approximately 14.2% up to 19%. We didn't want to add outside air because the mix of gases would disturb our other research on the composition and dynamics of our atmosphere. Gaie consulted with Dr. Sylvia Earle about her experience as a deep-sea diver, and together recommended locating pure, measurable oxygen in the form of liquid oxygen, the same form of oxygen used to fill hospital oxygen containers. Shortly after, a flotilla of trucks carrying liquid oxygen came out from Tucson; the liquid oxygen was converted to gas by vaporizing the liquid with a heat exchanger before passing it through a connection in the west lung to then spread throughout the entire sealed structure.

After examining his medical data, Roy recommended that we add to the mission rules a lower limit for oxygen of sixteen percent. Above this level only a few biospherians experienced early indications of high-altitude sickness—namely, difficulty sleeping at night without extra oxygen being pumped into their rooms. For these people, extra oxygen was provided by an apparatus in the Biosphere 2 lab.

Later, we again injected liquid oxygen once more so that scientists and technicians unaccustomed to our lower oxygen levels could go in and out of the Biosphere during the transition period between our mission and the next.

OUR ENDANGERED SPECIES ACT

In addition to the atmospheric challenges of carbon dioxide, oxygen, methane, and nitrous oxide, there were also ecosystem issues requiring our attention from the beginning. It was highly uncertain whether we could succeed in maintaining distinct biomes over time and whether our initial richness of species could be maintained. Some people warned us that Biosphere 2 might decline into a soup of algae as the higher plants and animals died off. Some leading ecologists even predicted a severe loss of species. About a year before closure, the two great founders of systems ecology, Eugene Odum and his brother Howard T. Odum (H.T.) visited Biosphere 2. H.T. laid a wager that within the first two years we'd lose eighty percent of our initial species, and that the remaining twenty percent will ignore our carefully crafted design for different biomes and organize themselves into something completely different. He didn't doubt that Biosphere 2 would work as a total system, but it would be nothing like the world we designed. Linda bet that losses would be less than twenty percent after two years.

Dr. Odum returned just before our first-year anniversary. With relief and great humor, he conceded that he had lost the bet. He was impressed that we were evidently holding on to sizable biological diversity and surprised to see that the biomes still showed very distinct and different characteristics. One factor he may not have fully appreciated is the sophistication of the environmental technologies that were developed to assist the biomes in maintaining proper temperature ranges, generate waves and currents, and produce rainfall. All of these factors were carefully adjusted to fit the needs of the ecosystems.

As early as 1966, the Odums unsuccessfully urged NASA to take a more ecological approach to designing life support systems. H.T. had long advocated taking a sealed greenhouse and literally tossing in plants, animals, soils, seed, and people as an experiment to demonstrate the power of ecological communities to organize themselves. In such an arrangement, he argued that there would probably be sizable extinctions before the system settled down. Our strategy actually did include a 'species-packing' approach; we included nearly 4,000 species of plants, animals, and uncounted microbial species. We did it this way mainly because no one knew which species would be lost or which would do well. The ecological designers tried to include a number of species in each biome that performed the same ecological task and role in food chains. This ensured that even if some species were lost, the system might remain relatively undisturbed, as other species filled in as backups.

Part of our species protection included occasional intervention to assist individuals and species that needed extra attention. An example was pruning back plants which were stealing the light from understory plants (such as morning glory vines), adjusting rainfall and temperature patterns to match the changing needs of the biomes as they matured, weeding out invasive plants, or as was the case with corals, removing macro-algae invading the outer rim of a colony's tissues and reducing the sunlight corals require. Yet most of the time, we let events take their course because it is quite natural that some species will do better than others in a new environment, and the focus of our observations is what Biosphere 2 itself was doing. Some losses were also a result of an ecosystem maturing. For example, the development of tree canopies meant that some early developing species would eventually be shaded out by slower-growing giants. A critical issue in intervention was whether the endangered organism was vital to the ecosystem because it was irreplaceable in the food web. In addition, some species received special attention because of their value to the humans, whether for their beauty or their utility.

Exact numbers of species extinctions would be revealed more accurately in our detailed resurvey during the transition between crews. During this resurvey of all the plants, animals, fungi, and microbiota, we will also document what is new— corals that had spawned producing new colonies, or seeds that were carried in with the soil that had sprouted, or plants that spread by establishing new seedlings, as examples. In addition, we were looking forward to learning in detail about which species wouldn't make it and/or at what point we would experience the evolution of new adaptations or speciation. We'd also take measurements of all the plants' dimensions to determine their rates of growth and how the original stock of carbon was distributed. For example, Gaie had only seen one species of hard corals (out of thirty-four) that was lost. We find this rather remarkable given the sensitivity of coral reefs and the notorious difficulties associated with their relocation. Losses would likely vary in different biomes.

In addition to the power of sophisticated sensors and analytic techniques to detect substances, there is the human organism, one of nature's most sensitive sensors. People can detect extremely small changes in temperature, humidity, and odor. A trained field ecologist is even more finely attuned to the many factors that give an ecosystem its characteristic feel. So, in a sense, we used a dual set of monitors to ensure that environmental conditions and life systems were healthy in Biosphere 2: the artificial intelligence of electronic sensors, chemical analyses, and human intelligence and sense organs.

The biospherians took regular walks through the wilderness biomes to check their overall health and the health of individual species, as did John Allen on the outside who would report his observations to Gaie daily. In many of the biomes there were organisms which served as 'biological indicators', a sort of early warning device of emerging problems. For example, in the coral reef the colors of the giant Pacific clams and the corals indicated whether or not the organisms were healthy. In other biomes, certain plants were checked to see if their buds had swelled to help us determine the best time to activate a dormant ecosystem like our savannah, thorn scrub, and coastal fog desert. We also kept our eyes on the dominant vegetation, such as the red mangroves in our marsh system, as a bellwether of system health.

OUR LEGACY TO THE FUTURE

We had been especially vigilant because of the many unknowns we faced in this initial shakedown of a man-made biosphere. With dozens, if not hundreds of potential disaster scenarios, we had to be continually alert: in a small closed system, things happened with amazing speed. This speed-up of natural processes represents a good deal of the scientific value of Biosphere 2. It's like a time microscope; you can see more processes in the same amount of time.

In its first two years, we experienced first-hand how this new Biosphere made a good start on its long journey of growth and development. It was far different, more thriving, and with much larger biomass than the Biosphere we entered in September of 1991. Our knowledge and practice as stewards would be passed on to succeeding crews, who in turn, continued to learn how biospheres work as a total system to provide the home for all life. We realized that this is the most important undertaking for humanity, and as always has been. We are part of a singular, interconnected, evolving life support system. Sir Albert Howard, the English soil scientist who sparked the revival of interest in organic farming, once said that the primary duty of those working the land is to maintain its fertility so they hand it over to future users in at least as good a condition as they found it. This had been the ideal motivating the research department as it set our standards, to ensure that we would hand over a more mature but still remarkably diverse and beautiful Biosphere.

Mark Nelson hoists a ripe papaya, one of the most successful fruit crops in Biosphere 2.

CHAPTER 4

Growing Your Own

"[Biosphere 2] is a paradigm of what man here on Earth has to face in years ahead. We have a burgeoning population, a decline of arable land, and waste heaps that are growing continuously. We have to find ways of producing more food in less space to feed more mouths, with less non-recyclable refuse.

– **Dr. Arthur W. Galston,** Professor of Botany, Yale University

IT IS ONE THING TO GROW SOME FOOD, but quite another to live entirely on the food you grow. All of us trained in the prototype agriculture systems we used during the pre-closure development phase, both in the research greenhouses and in Biosphere 2. Even though we ran nine week-long experiments in Biosphere 2 living off what the agriculture areas produced pre-1991 closure, it was not the real thing. Our experiments were brief—and the supermarket was still just a short drive away. Everything changed when the doors closed behind us on September 26, 1991. Our grocery store, pantry, and restaurant were now to be grown in an incredibly confined plot of earth. We had to learn to make it into fields of plenty or face a long and hungry two years. A group of educated, urban Americans and Europeans were about to become subsistence farmers.

INTENSIVE IS THE WORD

We called our farm the Intensive Agriculture Biome (IAB). It was about half an acre of land, including the domestic animal bay and a tropical orchard. Out of this half acre came virtually everything we ate, supplemented by some bananas, papayas, and coffee beans from the rainforest and some passion fruit from the savannah. Our farm provided food (fodder) for the animals as well, so it was also the source of our milk, eggs, and meat. Half an acre is a very small piece of land to feed eight people. Even the vegetarian diets of India or China need far more land per person than we had. To add to our problems, the intensive agriculture area was covered by a series of vaulted space frame arches and glass, which, even with the best of designs, cut out more than half the sunlight coming from outside. Finally, we were prohibited by our own EPA from using any toxic pesticides, herbicides, or chemical fertilizers. Under all these constraints, we had to produce a nutritionally balanced diet for eight—and sustain it for two years. We were allowed to build up a three-month reserve of food from farming Biosphere 2 before closure to tide us over the first winter. We were warned, "There are no indigenous Native Americans to tide you over like Mayflower settlers with a bounty of corn, pumpkins, cranberries, and turkey, and to teach you how to farm, so you'd best have something to start with." This will also be true of starting a colony on Mars.

From our previous agricultural trials in the research greenhouses and in Biosphere 2 before closure, we knew it could be done. The numbers showed it wouldn't be easy—even without serious losses of insects or disease we might only just make the target. We did not know how many hours a day we would have to work on the farm to survive either. Would farm work leave us no time for our other research? There was a delicate irony here: we lived in a $150 million-dollar facility with some of the most innovative technologies ever developed, yet the fundamentals of survival united us with almost every other human since Adam and Eve. Could this high-tech, information-age system be a model for future intensive, chemical-free systems?

In order to develop a sustainable system that was going to work, we had to take skills and knowledge not only from modern agricultural science, but also from ancient intensive farming systems. An old proverb states that the best fertilizers are the farmer's footsteps—that is, it's the constant attention of the farmer that makes for a good crop. We all dreamed of a lush cornucopia, a Garden of Eden, a life-sustaining oasis. These images are far removed from the realities of twentieth-century farming. These days, farmers sit in air-conditioned tractors applying chemicals to ground they will hardly tread, producing food that goes to the world's increasingly large urban population which neither knows nor seems to care how and where its food is produced. There was going to be many a biospherian footstep in the IAB trying to make the farm of our dreams and fill our plates with a satisfying quantity and variety of foods.

VARIETY IS THE SPICE OF LIFE

In developing the agriculture system for the Biosphere, we knew that variety was the key. If we managed to produce the proteins, calories, and fats necessary to keep us functioning but the food was dull and unappetizing, then life would become not unbearable, perhaps, but certainly uninspiring. With some amusement one day, several of us watched a TV news story about some NASA-sponsored research at Purdue University which explained the process of food production for future astronauts. They were attempting to engineer three species of plants—rice (grain), cowpeas (protein), and rape seed (oil)—in order to supply astronauts with a minimum, nutritionally complete, diet. This may look good in theory, if you ignore trace elements and some enzymes, but some basic points seem to have been missed. One is that with only three crops, a failure of any one of them means disaster. Another provides more important lessons we can offer to those involved with space exploration, though we are perhaps not the first to learn it: variety and diversity in cuisine is essential to human well-being. If you consider living off the planet for a long time—not just a few days or weeks in a space capsule where the effort is focused simply on achieving a short-term goal—then you must design a life style that can support you not just physiologically but psychologically as well.

It's not possible to exist on a diet where all of our 2,016 meals over the two years were composed of three plants. We already had more than eighty different crops in our Biosphere 2 agriculture and would probably add more in the years to come. Just as we cannot change our genetic make-up, we cannot expect humans to change their culture overnight or to survive only on proteins, carbohydrates, and fat. A healthy psychological and social life is inextricably linked to a rich and varied diet.

We hear much about wilderness biodiversity but preserving agrobiodiversity is also a major issue in the world. The diversity of crops in the IAB certainly made it just as pleasing to the eye as any of the more exotic wilderness biomes. On a spring morning we would pass fields of waving wheat, some golden-brown awaiting harvest, some green with seed heads just beginning to fill, others still looking like an overgrown grass lawn. Another beautiful sight was the 'potagerie', the vegetable patch, with a variety of plants from lettuces to chilies to squash. Then there were the sweet potato patches covering the ground with vines, sorghum fields with heavily laden heads, and white potatoes at several stages of growth. Heading along the north wall we ducked under banana bunches hanging over the pathways, with slim papaya fruit ripening to a golden glow on trees in their shadows. Looking over to the right, we could wave to a group of visitors peering into the windows in front of the rice that was just emerging from the mud. Alongside the paddies arched great stands of tropical lablab beans. Heading to the kitchen, we passed through the tropical orchard—a dark, green shadowy world where banana trees soared twenty feet towards the high ceiling. We saw taros (a tropical starchy tuber) with their giant fan-like leaves, and delicate citrus, fig, and guava trees which have staked out a position at middle height. Grapes, mint, and pineapples grow in the understory.

The crops chosen for the IAB had to be suitable for our tropical conditions that alternated with low light during the winter months. They also needed to mature quickly to ensure maximum use of the small space, resist disease, flourish under the reduced light levels beneath the structure, and be good to eat. In other words, we were looking for the 'stars —known, little known, and undiscovered— of the agricultural world. We started with a system based on sweet potatoes, white potatoes, and bananas for starch; wheat, sorghum, and rice for grain; peanuts and goat's milk to provide the fat in the diet; and a variety of peas and beans which supplied high-protein foods for our diet. Animal products, milk, eggs, and meat, supply only a very small portion of our food and dietary needs and so were largely used for special dishes (like ice cream) and holiday feasts. We also grow a large variety of seasonal vegetables as well as herbs for flavorings and fruits as sweeteners for desserts.

A HILL OF BEANS AND A PLOT OF POTATOES

Even before closure we experienced disease problems with many of the more common bean varieties. As we went into our first winter, most of the beans in the ground were a tropical variety called lablab beans which seemed to show some resistance to disease. We heard from Australian growers that they should flower around the time of the fall equinox as days began to get shorter. But the days and weeks rolled by with no sign of any flowers, and every day we worried a little more. We even tried hanging black plastic over them long before sunsets to trick them into thinking it was time to flower.

Sally and Jane nervously discussed what we could replace them with and how we could use their foliage. At least we'd get a solid supply of high-protein fodder for the animals if we had to cut down the plants, then over ten feet tall. Finally, on October 19, three weeks after closure during an early morning Saturday inspection tour, Sally crackled the good news over our radio channel: "Beautiful purple flowers on the lablabs!" Curiously enough, we also had a U.S. variety lablab planted on the south end of the IAB where they trailed down towards the basement, making use of the abundant sunlight that's otherwise wasted on the concrete wall. We called it the south foodfall. Lablab had been there since the previous winter. Within a day they were putting out beautiful white flowers like their Australian cousins. Their internal genetic program with flowering closely geared to day length and light had triggered simultaneously in both of them. The lablabs bore abundant flushes of all-important, high-protein beans for the next six months.

Sweet potatoes were, from the start, one of our mainstays, but we had much to learn about cultivating them under our unique conditions. In the first year, the crops of sweet potatoes were magnificent to behold—but only from the surface up. Their luxuriant leaf growth was balanced by a paucity of tubers underground. After consulting with Dr. Phil Dukes from the USDA vegetable labs in South Carolina, we came to the conclusion that our sweet potatoes had life too easy! Plentifully supplied with nutrients and water, they had no incentive to plan for the future by storing food in their tubers. We were told to make life harder on them. First, we were to 'dry shock' them until they wilted. Then, we were to reinforce the message by ruthlessly cutting back their long-running vines so they couldn't re-root all over the place. This forces them to send their food down to the main tubers. We were also advised to plant them twice as thickly to compensate for the reduced light. At first, all this advice was followed a little timidly. It was only after we harvested a few plots and found that the yields were best in the driest parts of the field that we gained confidence in the new approach.

Thankfully, sweet potatoes are highly resistant to most of the insect pests and diseases we experienced in Biosphere 2. Not so with our other starch staple, the white potato. Concerned about the possibility of an outbreak of potato virus, Sally learned the technique of growing white potatoes from tissue-culture clones. Although it was labor-intensive, it looked like we'd have a way to ensure virus-free potatoes in the future.

But in January 1992 we suffered an invasion by a pest we had never before seen—the tiny broad mite. First it took an inordinate interest in our yard-long beans, then moved on to an important field of white potatoes started from the virus-free tubers. Since we were unable to adequately control the minute broad mites, we had to greatly reduce the amount of white potatoes we were growing, increase our sweet potato plantings and look to other sources of starch such as taro and green bananas.

Sally divided our tiny farm into eighteen plots. The crops were rotated from plot to plot. In this way, we enriched the soil with legumes (nitrogen-fixing plants) and lessened the chance of harboring insect pests in the crops or soils. Most plots hosted three or four different crops over the course of the year. Sally alternated long day summer crops that did best in intense light and high temperatures like sweet potatoes, peanuts, squashes, and sorghum, with those that do best in cooler temperatures and can manage with lower winter light levels: wheat, carrots, cabbages, beets, white potatoes, peas. Other crops could manage in either season, although their winter harvest was slowed by lower temperatures and diminished light.

Sometimes it took a finer touch to juggle the two sets of crops—especially in the border seasons as we changed from one regime to the next. For example, if temperatures rose above eighty-five degrees, wheat pollination lowered, and premature heading occurred. So a typical winter temperature control was set between sixty-five to eighty-five degrees. Should temperatures fall below sixty-five degrees when summer rice was flowering, poor seed set could result. Sweet potatoes like higher temperatures, but should temperatures get above ninety degrees, sweet corn can develop empty heads and tomatoes and eggplants produce little fruit.

We were highly dependent on our computer controlled 'technosphere' to help us with this juggling act. The temperature and humidity regimes were controlled by large air handlers in the basement. Alarms would sound if temperatures above or below the safe range were reached. This gave us time to adjust the air handlers before any damage was done to the crops. The technosphere also took much of the tedium of watering out of our hands through the use of an automatic irrigation system which watered individual plots according to the needs of the different crops. Even with all of this technical aid, frequent checks of the crops were necessary to ensure that they were growing properly.

Overwatering made our soil sticky, and it drained poorly. This type of soil condition also causes more carbon dioxide to be released into the atmosphere. Under watering harmed the crops and made the ground as hard as concrete. We put soaker hoses down in problem areas, often on the edges of plots, where the automatic sprayer didn't evenly cover the ground. A week or so before harvests, the irrigation was usually turned off so that the field could be turned while minimizing soil compaction caused by the crew slogging through muddy soil. A programming error could (and on several occasions did) result in a group of biospherians being inadvertently soaked while weeding or harvesting a crop!

WORKING IN THE FIELDS

Planting and harvesting the rice was new to all of us. Several of us had watched it being done by village women in India, Nepal, or China, but actually doing it ourselves was an adventure. Wading around in the thick mud, we became more and more expert at transplanting the tiny rice seedlings. Since the paddies were right by the windows, the rice harvests and plantings were favorite events for the visitors, who could watch as crew members awkwardly tried to net the tilapia fish being raised in the paddies. Our rice paddies used a very space-efficient system. The fish ate azolla, a high-protein water fern that floats on the water surface, and in turn they fertilized the rice. The fast-growing azolla was also harvested as chicken feed. The rice took about four months to grow, so we got about three crops a year, although the winter crop was far poorer than the others. Timing here was all important. We learned to start new rice seedlings in the agriculture basement thirty to forty days in advance so they're ready to go straight into the paddies just after the next harvest.

The crew always enjoyed the peanut harvest. One or two people used shovels or pitch forks to gently dig up the peanut plants. The rest of the crew sat perched on upturned 5-gallon buckets picking the nuts from the roots of the plants and piling them into a bin. The greens were piled up and taken down to the basement to be dried in front of the air handlers for use as winter goat fodder.

Certain agricultural crews actually required a dress code. We discovered that delicate thinning and weeding jobs in between rows of plants must be done with bare feet so that the small seedlings do not get trampled by heavy work shoes. Sorghum, not a great favorite with anybody, required long-sleeve shirts to cover our arms and bandannas to cover our noses during harvest. This is because almost everyone had allergic reactions from contact with the leaves or spores of the heavy sorghum seed heads. Nearly all of us got into the habit of coming to IAB crews with a pair of pruning shears or a sickle tucked into our belts. Sally was rarely seen without a belt pouch stuffed full of essentials such as a lens for periodic spot-checking of insects (both pests and beneficials), a water-proof notebook, a pen for jotting down observations and crop yield data, and lengths of string for tying up the occasional straying tomato vines.

However efficient the automatic equipment we had, there was never any respite from the hard physical labor of farming. The system didn't wait for us to gradually adjust to living inside Biosphere 2. Fields ready to harvest were as insistent as a bedside alarm clock. Every day we didn't work in the fields was a day lost in the filling of our breadbasket, though we did just essential tasks, like supplying fodder and taking care of our domestic animals on the weekends.

We were not used to so much sheer physical work—digging soil, harvesting crops, and running threshing equipment. To make matters more difficult, we had to take on the work while adjusting to a lower-calorie, lower-fat diet than we'd experienced before. We soon learned to be as efficient as possible, working in large teams early in the day to accomplish heavy tasks like digging up potato fields, and allocating to later in the day the tasks that required less muscle power but more care and attention to detail. Almost everyone spent two hours after breakfast working in the agriculture system. Sally, as manager of the whole agriculture system, coordinated all the farm work. Different members of the crew took on various special responsibilities. Mark was in charge of fodder supply, the wastewater treatment system, the balcony planters, and the partly shaded south basement area that he turned into a sweet potato and papaya production area. Gaie took care of the orchard; Jane was in charge of the basic care of the field crops and the animals; Laser ran the compost system and raised worms in the basement; and Sally paid special attention to the vegetables in the potagerie.

DEMANDS OF RECYCLING

Not only did we have to raise all the food we needed for the two years, we had to do so while recycling all wastes, controlling pests and diseases without using toxic chemicals (going beyond organic standards since even plant-based sprays might impact our soils and water), and ensuring that our soils maintained their fertility. In order to maintain an agricultural system that was going to last over time—be truly sustainable—all nutrients taken out of the soil in the form of plant material has to be recycled. We couldn't use chemical fertilizers as it would introduce toxins into our closed system, nor should we need them. Our agriculture system, like all agriculture throughout history, should be able to maintain its fertility if we returned all the nutrients we took from it.

In many ways the animals were compost machines on legs. They happily ate many of the plant parts that humans can't digest and turned them into manure for composting. The water that washed down from the animal bay and the wastewater from the human habitat filled a series of holding tanks in the basement where the digestion of the nutrients they contain was continued. Next, the water was circulated through tank lagoons containing a constructed wetland system with a variety of fast-growing aquatic plants. Here the plant roots and their microbial helpers completed the conversion of waste into valuable nutrients. Many of the wastewater nutrients fuel luxuriant plant growth. The new growth was harvested for use as animal fodder, and, through the milk, eggs, and meat of our animals would be given back to us city-folk of Biosphere 2. The remaining nutrients were sent along with the water to the irrigation holding tanks to be delivered directly back to the agricultural soil they came from. As well, all our human wastes were sent to the constructed wetlands, so the nutrients from all our farm crops goes back into the recycling system and the cycle began all over again.

Composting and turning the fields caused carbon dioxide to be released into the atmosphere. During the winter months we had to be especially careful. Carbon dioxide levels rise then and we had to do everything we could to keep them from rising too high. This meant that compost material had to be shredded but stored dry during the winter months, ready for composting in the spring when carbon dioxide levels were lower. Our worm farms (we grew worms to replenish our agriculture soils because they are so good at breaking down organic waste matter to produce healthy productive soils) also had to be shut down due to the rising CO_2. We also turned to minimum-tillage strategies in our fields, sometimes just lightly rototilling the first few inches to make a seed bed rather than deeply digging the whole field. As in all aspects of life inside Biosphere 2, we had to take into consideration the effect of every action on the whole system.

PEST CONTROL

Our pest control program was another example of how our agriculture had to be compatible with the rest of this tightly sealed environment. It would have been literally suicidal to use toxic pesticides or herbicides inside Biosphere 2. With our small air volume and rapidly recycling water, we'd be breathing and drinking those chemical residues within days. We had to protect our crops by a variety of safe techniques such as using beneficial or good insects to control the 'baddies' (fourteen such allies were introduced before closure), using resistant crop varieties, using safe and biodegradable sprays, washing plants with water sprays, or even picking pests off by hand. Keeping the air humidity low also helped control many potential pests.

We were pioneering new techniques of biologically safe pest control in Biosphere 2. Some farmers in the outside world are using beneficial insects, which they can easily obtain from insectaries. But we had to sustain our populations of beneficial insects by providing food and habitat for them when they had eaten up most of the pests. Our chief enemy in the insect battles was the broad mite. These microscopic little mites loved the hot, humid conditions in the IAB. Our attempts at control varied from the marginally successful, such as lowering the humidity and introducing predatory mites, to the completely unsuccessful bright idea of blasting them off the leaves using a portable hair dryer!

The mite saga brought back memories of the fierce pest we battled in the months up to closure—the pill bug. This critter, with its shell-like exterior, is actually one of the few Crustacea (same biological family as crabs and shrimp) found in soils. Normally they function as detritivores, digesting dead organic materials from roots and plants. But occasionally they took a liking to the foliage, as we discovered in the spring and summer of 1991 when bean seedlings began to mysteriously disappear.

When we discovered the culprit, we began an intensive investigation of possible control measures. Pill bugs like moist soil, so one element of our farming had to be changed. Instead of keeping a mulch cover on the plots to keep them moist, we'd let them dry out a bit during the day to deprive the pill bugs of their cover. Although they were sometimes seen in the daytime, pill bugs are most active at night. They seem to boil out of the ground from particular underground spots where they congregate. We set out baited traps for them and even contemplated running pipes—cut in half and filled with water—as a kind of moat around plots planted with crops they particularly relished.

Then someone decided to try vacuum cleaners. After all, there were thousands of them, and since they seemed such a social lot we could start where they congregated. During several of the early week-long experiments, night crews of biospherians vacuumed bean plots by flashlight. Impressive numbers wound up in the bag, but this made little difference when we tried to get our soybeans started. The technique, ineffective as it was for controlling pill bugs, did at least provide a tasty snack for our chickens.

Our next tactic was to place plastic collars, sunk a couple of inches into the ground, around the young plants. The pill bugs must have laughed as they dug their way under our barriers. Then someone had the idea that rather than taking the pill bugs to the chickens, we could put the chickens in the field. So, in an effort that will be memorialized as one of the Truly Ludicrous Moments in Bug Control, we had chickens on tethers staked in the middle of our bean field, patrolling for pill bugs. We even made some portable chicken barriers so that we might set up a temporary chicken run in between crops. The modest rate of ingestion of pill bugs by the chickens, and the fact that we needed to get plots replanted immediately, ruled this out as a viable option.

Finally, a chance observation was made: Tomato plant prunings, if left on the ground and kept wet by our automatic sprinklers, were invariably thick with pill bugs. Eureka! We could put out tomato leaves as a kind of border around the vulnerable plots, thus drawing pill bugs away from the seedlings. It was a wonderful solution because not only did we keep pill bug numbers down, but we got a high-protein additive to the diet of our chickens.

The cockroaches also provided an extra snack for the chickens. Cockroaches are a species that clearly enjoyed life in Biosphere 2 as much as they enjoy it almost everywhere else. We introduced an exotic Surinam cockroach into the rainforest to help chew up dead leaf matter, and these stayed in their econiche. But the other, more common types managed to sneak in during construction. Their numbers began to proliferate, though at first they were mainly restricted to the wilderness biomes and basement. Soon they migrated to the kitchen and agriculture areas. To get rid of them we tried trapping them in buckets or mason jars baited with rotten banana peels. By the second year they were getting out of hand. These large roaches decided not to stop at merely breaking down dead plant matter, which is their rightful role in the food chain. They took a strong liking to nearly every type of living plant in the IAB. Tomatoes, sweet potato leaves, and the flowers and fruits of squash plants were a particular favorite. You can imagine her surprise (not to mention horror and revulsion!) when Sally ventured down with a flashlight to find herds of cockroaches grazing on rice seedlings and nearby sweet potatoes. They even made a go at eating green sorghum seeds by climbing up the eight-foot-high stems! We scrambled to harvest ripe papayas and figs before the cockroaches found them and burrowed their way in. To try and keep them under control, the Scientific Advisory Committee and our research consultants introduced halfway through year 2 a predatory wasp which parasitizes their eggs, as well as other beneficial insects. These added to our active research program on Integrated Pest Management (IPM), a vital part of sustainable farming as it offers an alternative to the use of toxic chemicals. Also, nightly vacuuming of the front and back kitchens became part of the day watch's duties. In the morning the vacuum cleaner is emptied in the chicken coops. The silver lining to this particular cloud was that the chickens liked the high-protein cockroaches and egg production increased. Additionally, both Sally and Linda noticed that in the agriculture and rainforest, ants and cockroaches were also becoming pollinators as they scrambled up and down plants carrying the pollen from plant to plant.

At one point, powdery mildew on our squash became a problem. Sally tried to control it by spraying the leaves with a potion made from our food-grade machine oil mixed with bicarbonates used in buffering the ocean. She read that this is a possible control for mildew and, sure enough, for some time it did a successful job. A small example of how we biospherians have to be quite resourceful and use whatever we can find to manage a problem.

CATCHING THE SUNFALL

Other clouds loomed. Well into our first winter, we became aware of a problem over which we have very little control. Sunlight, our most precious resource, already reduced by more than half by the glass and space frame structure was lessened even more by unusually heavy cloud cover. The normally high percentage of sunny days is what made us select Arizona as the location for Biosphere 2 in the first place, but in this year abnormal weather patterns worldwide (this was an El Niño year) caused excessive rain and flooding in Arizona and an unusually high number of cloudy days.

Each successive cloudy day brought rising carbon dioxide levels and more problems with the development of food crops and their susceptibility to disease. Every bit of sunshine was precious, so a change of thinking and process was begun, which stayed with us throughout the two-year experiment. We called it 'catching the sunfall.' This involved taking every possible space we could find in the agriculture area that received sunlight, no matter how small an amount, and catching the sun there with plants.

Mark was one of our chief 'sunfall catchers.' He started by cramming as many pots and plastic jugs full of soil as possible into the south basement area, using them for an extra crop of sweet potatoes. After that, he and Laser teamed up to create additional planting boxes using artificial lights that were no longer needed in the algae-scrubber tanks. They scrounged up wooden planks or cinder blocks to make the planting bed, hauled the soil in, and hooked up the lighting systems. Under these lights, red beets, white and sweet potatoes, squash, and even radishes flourished. Planting boxes sprouted in the IAB basement along the sunny south wall. There were even two in the plaza outside the IAB doors where we took our breaks. One day, when several of us were sitting and chatting over breakfast, Gaie told us that the previous night at three o'clock in the morning she ran into Mark who was gingerly moving a large lemon tree down the hallway on a dolly. Mark, subject at times to spells of insomnia, was working into the night to cram the balcony full of plants. Soon the balcony, which received some of the best light in the IAB, was home to thirty new papaya trees and 400 running feet of planters with lablab beans and sweet potatoes. It became a running joke that Mark would eventually plant up our personal rooms or put floating tubs out in the ocean to soak up more light. But it was also deadly serious; we needed more food producing space—especially if the low light levels continued.

That lemon tree which Mark dollied along began the transformation of the balcony overlooking the IAB. There we grew sugarcane and Leuceana trees after closure, hoping their fast growth would give us a boost in carbon dioxide management. But the sugar cane produced a lot of green stalk and not much sugar; so, Roy took over the balcony to grow beans. Impressed by how fast papaya trees come into bearing (nine months after planting you can harvest your first fruit) and with the quantity of fruit they produce (up to hundreds of pounds on a fully productive tree), Mark began to interplant papayas and bananas next to the beans on the balcony. By the time he was done, there were twenty-five more fruit trees on the balcony. There in the full sun we ripened papayas to use as fruit, while in the south basement area, with its poorer light, we harvested the papayas green as a vegetable.

From there, Mark continued his expansion by constructing a line of planters all around the front part of the balcony terrace. The light was excellent, and the beans and sweet potato vines could hang down over the railing to catch some of the light that presently fell unused. We organized bucket brigades to winch up soil from the IAB to fill the low containers and pots. We transplanted fig, lemon, and orange trees from the orchard, where they were doing very poorly in the low light, to the space around our café tables where we usually have our Friday dinners.

The process became a meditation on space and light. Suddenly it was possible to see new patterns and opportunities. One evening Mark discovered a new pet project. He was idly looking down on the IAB with Laser when they had another 'food flash'. "The stairwells! We can plant up the stairwells!" The IAB has four large stairways. The ones on the north were shaded by the wall of bananas and papayas, but the one on the south was hardly ever used and the one we did use could be boarded over. The two spots have excellent sunlight and add an additional 250 square feet of planting space! The project was rapidly put into motion with Laser engineering the main structure and everyone else helping to move soil and fill the planters.

DIALOGUE OF BIOSPHERES

By the second fall we really felt that we were coming to grips with the challenges of food production and were starting to understand exactly what it took to create a completely self-sustaining agricultural system. We also felt it was time to invite outside consultants to join us in this incredible venture—especially where we needed additional expert advice. So, Sally and Mark organized an agricultural workshop so we could share our data with others in the field and try to interest a top group of agriculturists to collaborate on long-term research just as we had done with the marine and terrestrial systems.

The key person we discovered was Dr. Dick Harwood of Michigan State University. He was recommended by Eldor Paul, the soil scientist from Michigan State who'd attended our April 1992 Closed Ecological Systems and Biospherics workshop. "Dick is the man you're looking for. He's a field agriculturist. Knows the science and is practical as well." Indeed, he was. He had served as coordinator of Asian programs for the international aid group, Winrock International;

directed the research program in organic farming for the Rodale Institute; and served as ecological consultant to the International Rice Research Institute (IRRI) in the Philippines when they decided to use an ecosystem approach rather than the high inputs, monocultural chemical path of the Green Revolution. Currently Dick holds the Chair of Sustainable Agriculture at that previous bastion of industrial agriculture, Michigan State University. He is also involved in the landmark National Academy of Sciences' reports on Sustainable and Alternative Agricultural Systems which challenged the current policy of subsidizing and encouraging expensive chemical agriculture with its attendant serious environmental pollution and high costs.

Dick in turn led us to Dr. Jim Litsinger, an entomologist with extensive tropical fieldwork behind him. Jim worked in the Pacific Islands during his Peace Corps days, then went to IRRI to consult on biological control strategies for rice and other tropical crops. From Winrock came Dr. Will Getz, an expert in small animals who consults to development projects in tropical regions of Africa.

Finally, in our agricultural workshop we have found our natural allies: experts who can see the relevance of Biosphere 2 to the accelerating shift all over the world away from chemically-based agricultural systems, and people with the practical skills to enable us to integrate what was being learned elsewhere. Dick was especially intrigued with the possibilities. An expert in the history of agriculture, he sees that we are approaching a watershed as we leave the fossil fuel subsidized agriculture behind and forge new types of sustainable systems. How to do so while preserving a wide biodiversity of crops, lowering labor requirements, and producing enough to remain economically viable was the challenge. Biosphere 2 was set up to work on that program while offering the research opportunity to study in detail how such a system operates and impacts the rest of a biosphere. As Dick told us: "the watchword in agriculture is becoming 'containment of nutrients.' What you have in Biosphere 2 goes beyond that—it gives us the chance to study the whole recycling even of those nutrients that are put into the air or drain through the soil." After working with us in a closed ecological system, Dick realized that what revolutionizes thinking is that there is no "away." You can't pretend to throw something "away" in the air or water, for it remains in the system and must be recycled if it isn't going to pollute and damage the living world. We are finally starting to see that this is also true in Earth's global biosphere.

The steps our new agriculture team was instituting to improve our production lifted our spirits during the dark, dark days of winter. Jim Litsinger immediately got onto the case of our continuing nemesis, the broad mites. He located four new predatory mites which were soon introduced on a scientific import. One radio station distinguished itself by trumpeting "Biosphere 2 Introduces 30,000 Mice." It was eventually forced to retract that arresting story—and presumably cancel the order for the Pied Piper. It was *mites*: 30,000 predator mites fit in two small test tubes! With the counsel of Will Getz, the research staff in the BRDC also started looking into animals such as Muscovy ducks, which might do well inside the Biosphere and be introduced for Mission Two.

Dick Harwood extended our soils research by working with colleagues at Michigan State University. An expert in Chinese farming, he thought the best parallel to Biosphere 2 was the intensive agriculture of Szechwan Province. Monsoonal conditions there parallel our reduced light. He showed us slides of a farm he'd visited there where nearly one hundred crops were raised on less than three acres, with a range of crops adapted to the conditions from the steep hillside slopes to the tilapia fish and rice cultivated in the ponds and paddies at the valley bottom. Dick began a worldwide search for other areas with similar light conditions. He also searched for shade-tolerant crops that could boost our productivity. The matches included monsoonal areas of the Philippines, Kenya, and the Ecuadorian rainforest. During transition we would further sculpt our IAB so that new crops from these areas could help us match the varying light conditions.

We could appreciate the importance of shade-tolerant plants. Our second winter was even darker than the first. It surprised meteorologists that there were two El Niño years in a row. Southern California and Arizona experienced record-setting rains and floods in January and February of 1993. Our light was twenty percent less than the poor levels of our first year! It was quite the shakedown test for our agriculture system!

GETTING THE HANG OF IT

Sally continued to perform miracles by juggling our food supplies. We had to dip into seed stocks to make it through, but even so, we felt successful: we didn't need to import any food during that long, dark winter of our second year.

Nestled in the foothills of the Santa Catalina mountains north of Tucson, Arizona, the 3.15 acres Biosphere 2 facility is the world's largest closed ecological system. Inside are tropical rainforest, savannah, desert, mangrove marsh, coral reef biomes, a half-acre farm, and human living area.

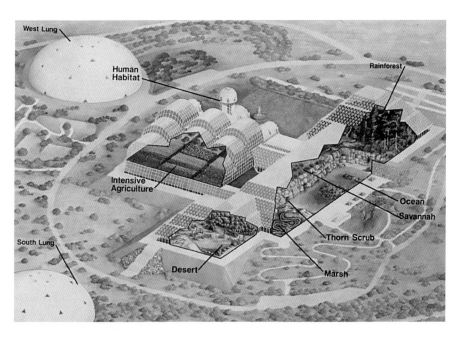

Schematic of Biosphere 2, built for one hundred years of operation, each of its biomes covered about half an acre and with deep soil/sediment tanks that were lined with stainless steel below ground and airtight space frame roofs above. The white geodesic domes covered its two variable volume "lungs" which allowed the facility's air to expand and contract without damaging the structure.

Biosphere 2's Mission Control housed a key part of the facility's "Nerve System," allowing outside scientists and IT specialists access to the data being collected.

Biosphere 2's operation required the expertise of many scientists and engineers. A sophisticated real-time data and sensor computer network, state-of-the-art for its time, facilitated the flow of information to and from the facility.

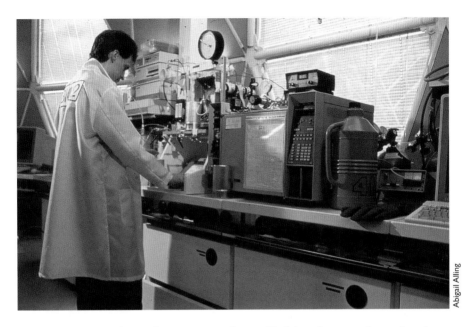

Taber in the Analytic Laboratory in Biosphere 2. The lab used no toxic chemicals and was equipped to make very detailed analyses of our air and water.

In addition to automatic sensors distributed around Biosphere 2, hand-collected samples of air, water, biomass, and soil were gathered for periodic analysis and to augment research studies. Here, air samples are collected in the savannah.

Fernando Monasterio

Biosphere 2's constructed coral reef included around three dozen coral species, of which only one was lost during the experiment. Its operation yielded important insights into the potential impacts of climate change on ocean health.

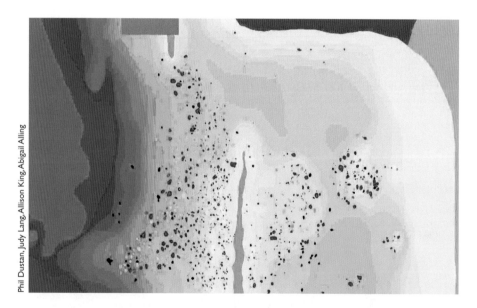

Phil Dustan, Judy Lang, Allison King, Abigail Alling

This map of the coral reef was done within two weeks of the end of the two-year closure. Living stony coral species are colored, with thin black lines representing seagrasses. Black colors represent dead colonies of all species. Depth was measured to generate a bathymetric contour map which was digitized and merged with the coral file to generate the final image.

D.P. Snyder

Several times a week, Gaie dove in the Biosphere 2 ocean to garden the corals and inspect the health of the reef.

Peter Menzel

The coral reef biome involved daily management and careful observation which included measurements of ocean water chemistry, weeding algae to remove nutrients, maintaining the pumps, wave machine, and other technical support systems.

A botanist in a new world, Linda measures inter-nodal lengths on a desert plant. Research conducted before, during, and after the closure experiment tracked the development of all the wilderness biomes.

In the rainforest, Linda reseeds planting pockets of the cloud forest mountain overlooking the lowland forest area. Fast-growing trees formed the initial canopy, protecting light-sensitive ones which will dominate the rainforest as it matures.

The desert was modeled on coastal fog deserts like that of Baja, California. Each biomic area includes a variety of ecosystems with characteristic soils, streams/waterfalls/pools, and vegetation.

The rainforest soon after closure, looking towards the cloud forest mountain. Water overflowed the pool at its top, creating a waterfall into Tiger Pond at its base and flowing through the meandering stream of the varzea, the low-lying floodplain rainforest.

The intensive half-acre farm had 18 plots, including rice paddies with rotating seasonal crops. The farm was one of the world's most productive, supplying over 80% of the biospherian diet for two years with more than 80 crops for food variety. .

The mangrove marsh biome was modeled on an Everglades, Florida estuary. Its zones went from freshwater marsh through five more ecosystems of increasing salinity, including oyster beds, white, black, and red mangroves. The mangrove marsh grew rapidly and studies confirmed its ecology closely resembled the natural areas where it was collected.

Finally, starting in March 1993, we were really beginning to see an improvement in our crop production. The sun finally came out and gave us the glorious, long, sunny Arizona days that had led us to choose this location for the Biosphere. Even some of the bean varieties that had been impossible to grow for a year and a half started to do well. The sixty new papaya trees planted during our sunfall harvesting campaign started to thrive, and hundreds of pounds of extra beets and sweet potatoes came in from our makeshift planters. As we suspected, it took the reality of being dependent on our agriculture to make successful farmers of us. What a difference when we understood at the stomach-level: if we didn't grow it, we didn't eat it. We had learned a lot from the experience, truly learning by doing, and we eagerly looked forward to the first transition phase where we could make alterations and improvements. Of course, as in all farming systems, new problems would continue to come up. Learning how to deal with them would be part of the fascinating process of creating a sustainable, high yield, and non-polluting agriculture system.

Sally Silverstone carefully weighs out the food for the cook of the day.

CHAPTER 5

Hunger and Resourcefulness

> "If it were up to the media, nobody would ever dare try anything new, exciting or adventurous, ever again. You mount an ambitious experiment at your sure and deadly peril. Because if everything does not go precisely as planned the very first time out, the media will crucify you—thoroughly and with obvious glee. Of course, the whole idea of an experiment is NOT to show that you already know all the answers but to find out what you DON'T know!"
>
> – **Brenda Forman,** International Space University and
> US Space Foundation, September, 1993

WE TOOK GREAT CARE to see that each member of the crew got an equal portion of food. We unanimously agreed that everyone should receive the same quantity regardless of age, sex, metabolism or work regime. Organizing it any other way would have been almost impossible, and probably cause a great deal of resentment. In the early months after closure, the cooked dishes were put out in large bowls on the counter top for people to come and help themselves in buffet fashion. However, as time went on and the crew got hungrier, there was a tendency for the early birds to overfill their plate so that anyone who was a little late for a meal was left with very short pickings. By the end of the first six months, the cooks were dividing the food into eight equal portions before serving it or using serving dishes where the division into eight portions was obvious.

Before the eight of us entered the Biosphere we got together and put on a satirical skit about how we envisioned life in Biosphere 2. One of the funniest scenes depicted the crew going down to the storeroom one at a time to steal food when hunger set in.

After closure, life began to imitate art, and this time it wasn't so funny: one or more biospherians were stealing ripe bananas from the basement storeroom. At first, Sally decided to ignore the thefts, figuring that the bunches of ripe and semi-ripe bananas were almost irresistible to a hungry stomach. However, over the year it slowly got out of hand. On one occasion sixteen ripe bananas that she earmarked for the day's allotment of food disappeared before she could harvest them and take them to the kitchen. It was then that she decided to hang the ripening bananas in a locked storeroom. The processing room freezer also became an overwhelming temptation to at least one member of the crew. When Sally discovered that her precious frozen food supply (bananas and other foods) had been hacked into by a person or persons unknown, she decided that this too had to be locked.

MAKING PLANTS TASTE GOOD

You know that they make bread with wheat. And you probably have some idea what a field of wheat looks like. But do you have any idea how those plants become bread? Most people have at best a vague notion of how such foods are actually processed to make them edible, but in the Biosphere, we needed more than a vague notion. Planting, growing, and harvesting were only half the story: we had to make food out of what we harvested.

Finding suitable equipment wasn't easy. We operated on too small a scale to use machinery designed for large farms or food processing plants. On the other hand, we had to minimize our labor with some kind of machinery, and needed equipment for a wide variety of crops. Some, like sugar cane, are not common in the United States, so it took quite a search before we found a good simple sugar cane press for squeezing the sweet liquid from the stalks.

Wheat, sorghum, and rice—our principal grains—can't be eaten without considerable processing. When we harvested our wheat, we cut handfuls of the sheaves with a pruning shear or sickle, stacked them in bundles, and laid them into large carrying buckets with the seed heads pointed out. Then we carried them to the basement of the IAB, where we kept the equipment to make the stuff edible.

After drying the wheat for several days in the large drying ovens, we started threshing. The threshing machine was a little small for the job, and could be dangerous if operated by an inattentive biospherian. Soon after closure we got a vivid demonstration of the dangers. Jane assigned Roy to thresh rice down in the basement of the IAB. He got the machine started, and began feeding rice into it, but yelled for help when he couldn't get it to work properly. Jane, working nearby, charged into the room.

"Look! If the stalks build up inside the machine, it won't be able to thresh the rice. That's the problem," she said, suspecting that Roy was looking for an excuse to pass the job on to someone else. She put her hand into the slow moving barrel to pull out the stalks. "Oh God! Roy, turn off the machine—my hand is stuck! Turn off the machine!"

After what seemed like an eternity to Jane, Roy turned off the power and Jane lifted her hand. The top part of her middle finger was nearly severed. Roy took her to the medical room, where he and Taber reattached it. After it was clear the graft was unsuccessful, Roy finally decided that Jane should have the benefit of specialized hand surgery and insisted that she go out of the Biosphere to a hospital in nearby Tucson. Within five hours, Jane returned to Biosphere 2 through the airlock. The operation was successful, but there were many weeks to follow before she could resume work. It was a lesson for us all about being vigilant and careful in our myriads of technics which augmented all of our needs as well as the life support systems. We were not living in 'nature' with giant predators or illnesses to avoid, we were living in a highly advanced technologically man-made life support system.

After that experience, we tried several alternate ways of threshing the grain. One was to lay all the wheat on a tarpaulin and have the entire team stomp on it with bare feet. This was fun initially, but we soon realized that it took more time and was less efficient in removing the grain from the hull. Concerned about loss of time, Sally and Linda—the most efficient threshers—took over the threshing work. They soon developed their own methods of making the machine work. Threshing away in the dimly lit basement was rather like digging for gold in a deep mine—the pleasure came in watching the loot pile up in the buckets and then weighing the final grain harvest.

After threshing, the grain was winnowed to blow out any remaining husks and then ground into a coarse whole-wheat flour ready to make breads, pie crusts, cakes, pancakes, crêpes, or porridge. We stored the straw from the wheat for use as domestic animal bedding. This ensured that any grain that escaped threshing was easily found by the chickens or the goats.

Experimentation led to discovering the best way to process dry beans. Eventually we hit on the idea of stuffing the beans into a burlap sack and pounding it with a rubber mallet. This breaks open the dry shells and then the beans could be winnowed out, ready to store or cook. This was a particularly good job for someone who needed to take out his aggression on an inanimate object: banging on the sack, under the right circumstances, was a very therapeutic activity!

We were less successful with the dehulling of rice. Sally has clear memories of the village women of Bihar, India spending a considerable amount of their waking day pounding the rice with a large stick in huge wooden pestles to remove the hulls and then winnowing the grain and hull mixture by throwing it up into the wind over large woven mats. Clearly we didn't have time to do this in Biosphere 2. Eventually, we obtained a small Japanese dehulling machine, but it continually jammed if even the smallest piece of straw got caught in it. After many frustrating hours spent taking the machine apart to unjam it, Sally developed a method of operating the machine with one hand while holding it together with the other. This way when it jammed it was easier to open and clear the machine out.

Once dug up out of the ground, sweet potatoes have to be cured by leaving them in a cool, dry place for several days before they could be stored. When the fields were dug up, everyone tried to guess how much we harvested. Between the eight of us, we ate approximately ten pounds of starch a day, so the big question was always how long a particular crop would last. A good harvest meant sighs of relief all round, and maybe a few extra sweet potato pies. A poor harvest often sent Sally on an inspection tour of the banana trees and taro plots to calculate how much starch we could get by eating them until the next sweet potato crop was ready.

Fruits and vegetables were harvested daily and used as part of the day's cooking allowance. For example, we cut banana bunches from the tree when they filled out and a few bananas showed signs of yellowing. Later, the rest of the banana tree was cut down and composted, its leaves carted to the animal bay for fodder. The green banana bunches were hung in the basement and left to ripen. Since bananas were the main source of sweetness in our diet and they by far out produced our ripe papayas, figs, and guavas, they were carefully rationed so we could have some every day. When a whole bunch of bananas ripened at once, they were broken into small pieces and frozen in three-pound bags for daily cooking allotments.

COOKING AND EATING

Once every eight days, each member of the crew cooked for everyone. The cook's 'tour of duty' started with the evening meal and continued over breakfast and lunch the next day. This way soups slowly cooked overnight in ceramic crock pots, and cooks organized themselves so that the bulk of the cooking was done the previous afternoon, leaving them free to concentrate on other tasks the next morning.

Sally carefully weighed out the day's allotments of fresh foods for each cook. Dry foods, such as beans and flour, were weighed out on a weekly basis and left in the kitchen for the cooks to take their allotments. A board in our back kitchen displayed the changes in daily allotments of starches, flour, and beans, depending on what was available from recent harvests. This helped us keep very accurate records of the amount of food that was divided up. Sally then entered this information into our computer database that was programmed to determine the amount of nutrients we were eating. This ensured that we were keeping above the recommended intake levels of calories, proteins, and fats.

Most Americans have to watch carefully to make sure they don't eat too much fat, but our diet was only ten to fifteen percent fat—the low end of the range. Peanuts supplied most of this fat, though a small amount came from milk, eggs, and meat.

Feasting when we could became an important part of biospherian life. The joy of sitting back and stuffing the belly was a welcome contrast to the daily experience of being hungry after every meal. Each feast had to be thought about well in advance so Sally could save special goodies like milk, eggs, figs, and papayas. By the time a feast came around, we could really prepare a spread.

Coffee was such a precious commodity that decisions about its consumption were a matter of some gravity. Usually we waited until we harvested enough coffee beans for eight cups, then decided when we would drink them. Linda, whose love of good coffee led her to be particularly interested in its production, became the doyenne of coffee affairs. She announces how many 'coffee pots' worth of coffee we have in storage and gives projections telling us when we'd get our next harvest. Then afterwards, we'd seriously discuss exactly when and for which special occasion the next pot of coffee would be brewed. Linda also undertook the peeling and roasting of the beans. The night before a 'coffee breakfast' the smell of freshly roasted coffee deliciously permeated the habitat.

Each member of the crew had a different cooking style. Some cooks spread their ingredients out all over the kitchen counters and then spent a long time pondering how to divide them up into various dishes. Others first washed and chopped everything and then played it by ear, creating dishes as they went along. Some cooks preferred to make a heavy dinner, which meant that lunch was a little lighter. Others arranged things the other way around. The trick was to have enough for all three meals.

Breakfast was very important, especially since we did most of our heavy agricultural work in the morning. We unanimously agreed that breakfast should always start with a large bowl of porridge sweetened with bananas and other fruit when available. Some biospherians occasionally blended in cooked sweet potatoes, green bananas, green papayas, yams, or plantains. Side dishes with the porridge varied.

Either the porridge was accompanied by a sweet bread (for example, a sweet potato loaf with banana and figs on top), baked sweet potatoes, beans with chilies, or a small glass of banana milkshake, made, of course, with our sweet African pygmy goats' milk.

A typical lunch menu might have included a sweet potato and vegetable soup, sautéed vegetables, salad, sliced beets, and wheat rolls. Most evening dinners were vegetarian. A typical dinner could include a bean and potato burrito with hot sauce, steamed vegetables, salad, and banana ice cream. For Sunday dinners we enjoyed a small serving of fish or meat from the chickens, goats, or pigs. Aside from the three meals a day, there was a mid-morning daily snack of roasted peanuts and mint tea.

The cook of the day served dinners between 6:30 and 7:00 PM Breakfasts were served at 8:00 AM on weekdays and a leisurely 9:00 on weekends, while lunch was served daily at 12:30. Breakfast was always held in the dining room, which has a view of the upper savannah. As we ate, Sally called the morning meetings to order and each member of the crew reported what they'd be doing that day. At the end of the meeting any remaining porridge was rationed out by the cook. Lunch was also held in the dining room, but dinner was a moveable feast. The balcony over-looking the IAB became a favorite place to eat in the evening. This area had a café like atmosphere and became known as the Café Visionaire. We decided to hold our Friday night dinners there. Everyone enjoyed sitting back at the end of a long week to witness the spectacle of the sunset over the Santa Catalina mountains and our beautiful farm through the space frame as we ate. The Café Visionaire was also the location of some of our holiday feasts and special breakfasts. But lunch there was out of the question: during the hot part of the day temperatures frequently rose above 110 degrees.

Even though we increased the amount of calories per person after the first six months, and the weight loss stopped, it was clear that a calorie-restricted, low-fat diet without high-sugar snacks took a lot of getting used to psychologically—especially for a crew who had been accustomed to the variety of sweets and munch-ies available in an American or European supermarket. The only snack we had was after our two hours of agricultural work. In the first year, break time snacks were either a ripe banana or roasted peanuts. In the second year all the ripe bananas were needed for cooking, so our snack consisted of a handful of roasted peanuts and, occasionally, some sugar cane stalks to chew on. Other than that, food was not available between meals.

Biospherian adaptations to the food situation varied. Mark came to be known as the human goat, because he happily snacked on anything green that he could find to chew in the agriculture area. His favorite snack was the sweet leaves of our fennel bushes which he described as 'biospherian chewing gum'. It took Sally, who was in charge of the herbs, many months before she realized that the mysterious grazer of her fennel plants was a human. Both Mark and Sally, the two iron stomachs of the crew, took to chewing empty peanut shells or even the occasional ripe banana skin to supply at least the illusion that food was being taken in.

Sally squirreled, saving little pieces of this and that from her meal plates and putting them away in the freezer for a time when she was really desperate for a snack. She actually rarely delved into her store—it was more as if she were comforted by the knowledge that it was there. Other members of the crew save parts of their meal to eat later, especially if they knew they were going to have a late night or an especially demanding task to complete. Mark often saved his dessert for late night snacks while Taber, Laser, and Gaie ate everything at once, rarely saving anything for later.

Probably because we were mostly on a vegetarian diet and our ability to digest meat had decreased, some of the crew had problems with gas attacks after a heavy meat feast. Linda and Jane seemed to have greater problems with eating meat. By the second year, they decided to decline even the small amounts of meat we had available, so cooks prepared alternative vegetarian dishes for them on our rare meat nights.

All the biospherians had to learn to be adaptable and creative cooks, since the types of food available varied greatly during the year. Sometimes the main starch was white or sweet potato, other times it was plantain, green banana, or taro and eggplants. In the winter we had very large harvests of root vegetables such as carrots and beets.

Beets consistently did well in the IAB. Consequently, we often have more of them available to eat than we'd normally have included in the diet. Our Russian visitors always laughed when we told them how much borscht (beet soup) we ate. Beets froze fairly well, but tasted much better fresh. In response, the crew devised all sorts of ways of cooking and serving beets. We had a dozen variations of borscht, beet casseroles, and beet salads. Someone even tried to put beets in the porridge, but this was vetoed by the rest of the crew. Gaie invented a surprising and delicious cold beet juice with lemon and chili. Laser made a beet, sweet potato, and banana milkshake for dessert one night and tried to pass if off as raspberry sherbet. We at first joked that it'd be a good idea to make beet wine, but this never happened, though later we made a beet whiskey which was truly appalling.

As the experiment progressed, we became quite good cooks. This was one area where intense peer pressure worked near miracles: even the worst of our pre-closure cooks became reasonably competent. Each cook developed specialties. Laser was famous for his pastries and Belgian crêpes which became a favorite for feast and birthday breakfasts. Gaie made superb ice cream and buns. Taber cooked up individual pies and Mexican pastries in all sorts of strange shapes and created colorful and artistically arranged individual plates. Mark brewed up delicious vegetable soups, containing everything edible he could lay his hands on. Jane invented a wonderful posole recipe that was ideal for using up miscellaneous parts of the pig. Linda was known for her soups, including her cheeseless 'cheddar cheese' soup. Roy made a superb chutney and Sally became chief birthday cake baker, resident brewer of alcoholic beverages, and maker of specialty items such as goat cheese, cream cheese, and pork sausage. Eventually, Sally's reputation grew on the outside, until she was asked to do her cookbook, *Eating In: From the Field to the Kitchen in Biosphere 2*—which has been a great success.

Because of the intense anticipation everyone has for each meal, cooks became extremely careful. After a couple of incidents in the beginning months it was unheard of for a dish to be burnt. This was especially important not only because of food loss, but because the smell and smoke linger for a long time in a closed system. There were, of course, the failed experiments. For example, one cook tried to perfect a cold green soup for lunches by blending up Swiss chard and beet greens. After two tries, we all agreed that this was one experiment we didn't want to be subjected to. Another time Gaie and Sally got it in their heads to cook up marsh canna roots. They carefully washed and scrubbed the slimy mess that they extracted from the bottom of the marsh and then boiled and seasoned it. This turned out to be a disaster since the roots are almost entirely composed of fiber. Even hunger has its limits!

We continuously experimented with different types of food. One of our biggest challenges was making taro edible. Several members of the crew were sensitive to the oxalic acid contained in the root—it stung their tongues and upset their stomachs. It was certainly not a taste we were used to. We discovered that by peeling the outer layers of the root, boiling it several times and for several hours each time, discarding the water and then mixing it with other foods, we could make it palatable. Gaie hit on the best idea for preparing it: after doing all of the above, she then baked it for three hours until it became a crispy, golden-brown potato-like patty.

Hiding bland foods in other dishes became second nature. For example, we had plenty of green papayas harvested from trees growing in low-light conditions. They took too long to ripen, but they proved to be an excellent starch supplement. They were quite tasteless but useful in thickening up soups and sauces, and in small quantities are almost undetectable in a hot, morning porridge.

In the winter when vegetables were in short supply, we tried adding sweet potato greens to the daily diet. While these are a staple food in some cultures, some members of the crew had a hard time with the taste. Some cooks refused or 'forgot' to serve them, while others would happily gulp them down along with everything else, appreciating them as stomach fillers. But when other harvests kicked in no one complained when they were dropped from the menu. This left more sweet potato greens for the goats, who, so far as we know, never complained about them.

Plantain presented a problem at first as no one was really experienced in cooking it, but we soon learned to bake it and mix it into stews and stir fries. At first we didn't know what to do with long, thin, sweet potato tubers (some as small as a pencil diameter). We finally discovered that, when carefully baked, they come out with a tasty, crunchy texture. Potato chips!

But gradually our tastes began to change; our cravings for chocolate and junk food decreased. We found new ways of bringing out the best qualities of vegetables and fruits. On the whole, by the second year the crew agreed that we had really excellent food, and at times we all contemplated the fact that we may never eat so well again. After all, we knew exactly where our food came from and there were no preservatives or extra sugars and fats thrown in. Anyone who grows their own organic vegetables knows that the taste is marvelous and nothing like industrially grown foods. This was true for us; we ate from, and lived with our beautiful and bounteous garden which never had chemical sprays or fertilizers. Over time our vegetables and fruits grew sweeter to our taste as we were weaned from refined sugars and sugar additives. Ours was truly a natural and organic diet. The only problem was that there never seemed to be enough of it. We never learned who'd been stealing the food.

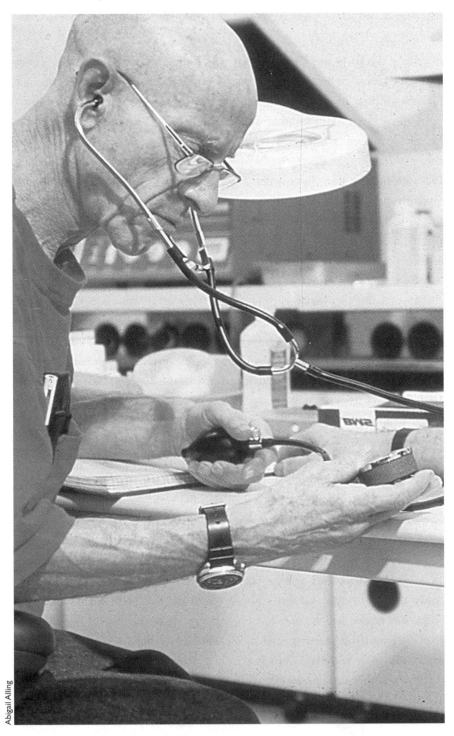

Maintaining health and vitality for the Biosphere 2 crew was top priority.

CHAPTER 6

The Doctor Is In

"I do feel that from many fields—biology, genetics, and even human health – many pieces of information will come out of their records that will help those sciences right here on Earth."

– **Richard Evans Schultes, PhD., FMLS.,** Professor of Biology and Director of the Botany Museum, Harvard University

In Biosphere 2, the doctor was always in. Every eight weeks, each biospherian was scheduled to meet with Dr. Roy for a full medical checkup which included routine checks of weight, pulse rate, blood pressure, body temperature, eyes, ears, throat, and reflexes. In addition, Roy took samples of urine and blood for chemical analysis. This way he could track our cholesterol and red and white blood cell counts. Every few months, Roy took a Polaroid photograph of each biospherian without clothes on as a documentary record of the physiological changes that occurred with our unique calorie-restricted diet and associated weight loss. Every six months, Roy ran an additional series of exams which included an x-ray of the chest and an electrocardiogram to check heart condition.

The Biosphere 2 medical facility was designed to minimize byproducts that cause environmental pollution, which would contaminate our tightly sealed world, so no chemicals that could harm the air and water systems were used for tests. We also tried to minimize the time required for medical diagnosis and treatment by using state-of-the-art technologies. For example, Kodak donated one of their Vitek systems, a device for analyzing bacteria. The machine was originally developed for use on NASA spacecraft, to identify the type of bacteria responsible for common infections. Indeed, some of what we were learning while operating our medical facility might be of value for remote human habitations such as in Antarctica or on a space station or lunar colony.

Roy was a graduate of the California Institute of Technology and the University of Chicago Medical School. He had been a professor of pathology at the University of California Los Angeles as a researcher in genetics and nutrition for over thirty years. When he decided to join the first two-year crew, he wondered if he really wanted to become a general practitioner, a 'country doctor' as he put it. But in the end, he took refresher courses in general medicine to get licensed as a medical practitioner in the state of Arizona so that he could provide a full range of services for the other seven crew members for the two years. These courses covered dentistry, general health, prescription of pharmaceuticals, and gynecology. Special care was taken to prepare for possible accidents, allergies, and infections. SBV sent Taber, his assistant, through a special, intensive six-month training program in general medicine—including dentistry—at the University of Arizona Medical School. He and Roy also completed a program in microbial medicine at UCLA so that if Roy required assistance, or was the patient himself, Taber could carry out the necessary procedures.

Roy's initial concerns about becoming our country doctor and being weighed down by endless complaints from sick biospherians proved groundless. As it turned out, we rarely got sick. Because we were removed from contact with germs in the outside air, what we went in with on September 26 is what stayed with us for the two years. Even our fears that we might pass around a few 'bugs' early on in the experiment were unfounded, perhaps because we'd been working and eating together so closely for months before full closure.

After we started to import and export research samples in July of 1992, we noticed that the biospherians who entered the airlock sometimes become slightly ill two to four hours later. These illnesses seem to match those on the outside. It seemed like we're highly susceptible to infections on the outside, and we began to wonder about what would happen when we emerged in September of 1993. It is possible that we'd become less immune to the common colds and flus because we had been living in such an isolated environment.

Since the Biosphere is full of plants producing an array of biochemicals and pollen, we thought allergies might become a problem. If some of these irritants had not been taken out of the air by the filters in our air handling units, the allergy problem could have been much worse. Jane was treated for her allergic sensitivities before closure, and Linda received a series of shots for her allergies from Roy during the two-year period. Even so, she still suffered a bit when the savannah grasses were in flower.

Some of the other medical problems tackled during the two years included sporadic allergy attacks, mild episodes of insomnia, stomach-aches (especially after early feasts when we hadn't learned to restrain ourselves in front of a heavily laden table of food), and accidents. Roy handled all our health needs with one exception: Jane had to leave Biosphere 2 for five hours after the threshing machine accident that severed the tip of one finger. Her finger had to be surgically closed at the University Medical Center in Tucson. Other minor accidents included the odd sickle cut and acacia or cactus thorns embedded in feet. Minor remedies such as band-aids and aspirin were kept in a little metal tray just inside Roy's medical office.

Roy's primary research interest covers the effects of nutrition on human physiology. Many times he had said that destiny played a hand in picking him as the doctor for the first team of biospherians because one of the most interesting medical aspects of our two years inside Biosphere 2 turned out to be our diet. For over thirty years Roy had been studying the effects of a high-nutrient, low-calorie diet on mice. He found a variety of physiological effects including lowered cholesterol levels, lowered blood pressure, an enhanced immune system, a dramatic slowdown of aging, and an increase in their life span. In sum, these mice were healthier than the mice that were eating a high-calorie diet.

A low-calorie, nutrient-dense diet was exactly what we wound up eating inside Biosphere 2. So, for the first time, research on the health implications of a human group eating this type of diet could be studied—and in a situation where cheating was impossible. Roy could be absolutely sure that what we produced is what we actually consumed—there could be no quick runs to the convenience store.

In the first few months we averaged about 1,700 calories per person per day. This was quickly raised to 2,000 and then maintained around 2,200-2,300 calories. It was nutrient-dense in that we ate grains, beans, and lots of vegetables—sugars and refined foods were non-existent, and fats minimal. The diet contains plenty of protein and enough of the other nutrients for good health. But there was a definite limit on the number of energy-producing calories we had—and a whole lot less than we're accustomed to. Each biospherian responded differently to the diet. Initially, over the first six months or so, we lost between eighteen and fifty pounds each. Mark remembers around Christmas of the first year looking at his weight on the scale and projecting that if that rate of weight loss continued, he'd weigh minus ninety pounds at the two-year point! But during the course of the experiment, after we adapted to our new diet, most of us either maintained our new weight, or managed to gain back some of what we had lost. Others never fully adjusted and were either losing weight or staying at a border-line level. Roy continued to assure us not to worry when we commented on our baggy pants and loose shirts because our overall health had actually improved by the combination of our diet and the superb freshness and quality of our organically grown food.

Roy was intrigued by the wide variation in our physiological responses. We ranged in age from twenty-nine to sixty-seven, differed in ethnic backgrounds, and were a mixture of metabolic types from lean and thin to stocky and chunky. But each ate the exact same foods and number of calories during the two-year period.

By the time two months had passed all eight of us had our first medical check, and we were all astonished by the results. We started with normal, healthy levels of cholesterol averaging just under 200. Now we were averaging less than 125—the levels normally only seen in young children with their years of fast food and high fat consumption ahead of them. These low cholesterol levels lasted for the duration of the experiment.

Roy the scientist was happy at the research opportunity. Roy the person experienced hunger like the rest of the crew. Hunger became our constant companion, always there to struggle against. Part of it was the painful adaptation from our high-calorie, high-fat pre-closure diets. No doubt part of it was adjusting to the reality mankind has lived with for most of our evolutionary history. In between meals we were almost always hungry. We came promptly and eagerly to every meal, motivated both by hunger and the desire to get a fair share of the food served. We often left meals, except for feast celebrations, still somewhat unsatisfied. This is part of our biological heritage—and unfortunately a reality for many people in the world today. Unlike many of the people in developing countries who are both undernourished and malnourished, our diet is dense in nutritive value. Still, being so restricted in calorie intake for so long a time was a new experience for us.

This was the first study of humans on this type of diet, and Roy published some of his findings in the *Proceedings of the National Academy of Sciences* in December 1992. Since we had shown the exact same short-term physiological changes as the laboratory animals, it was intriguing to speculate on whether staying with such a diet lead to slower aging and life extension like Roy found with the mice. We like to think that we may have grown younger instead of older during our two years in Biosphere 2, but there was no way to test that.

Many of the biospherians noticed that out of necessity they were changing old behavior patterns. Frenetic activity was out of the question on our restricted diet. No one ever heard anyone run up the stairs, and for the first time in years, many of us found ourselves taking a nap during the midday siesta period. The finite supply of food definitely curtailed useless action. There was no need to do anything in a hurry, nor could we afford to burn calories on non-essential activities. On the other hand, we were not sedentary. Most biospherians work two to three hours a day in the agriculture area, often doing manual jobs like planting or harvesting, as well as additional hours in their own particular management areas. Most of us found it hard to believe we could sustain active lives on such a reduced diet.

There were some other fascinating side-effects of our diet. For a while friends and family commented on how pale we were becoming. This was not surprising, since we were deprived of ultraviolet radiation which is responsible for tanning skin. But then we noticed that our skin—especially the palms of our hands—began turning more and more orange! This was a result of taking in high levels of beta carotene, a nutrient found in red beets, sweet potatoes, and carrots. Since these three vegetables were always in good supply in our pantry, we would walk out of Biosphere 2 with a markedly orange complexion.

Gaie and Roy were the only ones who had dental problems. About six months into the project, Gaie noticed a filling had fallen out of one of her teeth, and about a year later Roy had a similar problem. Luckily, Roy stocked a dental paste for emergency fillings. Roy's emergency dental paste has two ingredients which harden when mixed together. The hardening takes only two minutes, so it has to be applied quickly. Roy successfully patched Gaie's missing filling and Taber did Roy's tooth. This was a temporary fix—Roy cautioned that the long-term solution will have to be found when we exit.

DECLINING OXYGEN —
THE SLOWEST MOUNTAIN CLIMB IN HISTORY

The unexpected decline in atmospheric oxygen prompted another kind of medical research. It declined from 20.9%—the same as in Earth's atmosphere at sea level— to approximately fourteen percent. When the oxygen levels in the atmosphere reached about 16%, the biospherians began to experience symptoms of high-altitude sickness (headaches, shortness of breath, trouble sleeping). Roy contacted four scientists to help figure whether the biospherians would adapt to declining levels of oxygen in the atmosphere. Dr. Harvey Meislin, our main consultant at the University of Arizona Medical School, Dr. Peter Hackett of the University of Alaska, Anchorage, and Dr. John West, of the University of California, San Diego, thought that humans could adapt to lowered oxygen levels corresponding to those found at altitudes of up to 18,000 feet above sea level. They also warned Roy that there would be an appreciable difference in individual adaptation, and that it might not be a straightforward process because the decline of oxygen we were experiencing was occurring over a long period of time. A fourth consultant, Dr. Igor Gamow, a professor of chemical engineering and sports physiology from the University of Colorado, predicted that we wouldn't adapt. In his opinion it was not only the diminishing amount of oxygen in the air that was critical to acclimatization, but also the diminishing air pressure of high altitudes. For us, the air pressure remained the same during our two years.

During the last year of oxygen decline, Roy closely monitored all of us. He sent blood samples to several different labs to track hemoglobin and red blood cell numbers. These are two factors which normally increase with adaptation to lowered oxygen; they allow the bloodstream to carry more oxygen to the cells of the body. Roy and Taber initiated a program to track our physiologic changes with a standard set of exercises. Every few weeks we all went in turn to the analytical lab where a stationary bike was set up. We rode an exercise bike for two minutes, keeping the exertion gauge between 600-700 calories burned per hour. Meanwhile a forefinger was inserted into a device which measured both heartbeat and the level of dissolved oxygen in the bloodstream. These were monitored for a couple of minutes before we began bicycling, and for four or five minutes afterwards until the heartbeat slows down and stabilizes. Taber also hooked up a tube to a gas chromatograph in the analytical lab. By having us expel most of the air in our lungs and then blow through the tube, he's able to measure the oxygen content in that last portion of lung capacity. By the time of the oxygen injection in mid-January 1993, our oxygen levels had declined to around fourteen percent. This level corresponds to an eleva-tion of 13,500 to 14,000 feet. Some degree of acclimatization (a small increase in red blood count) was measurable at this point. In other words, it was likely that we were slowly adapting, while we continued to show the symptoms associated with oxygen deprivation. The worst symptom was a decline in work capacity. This was difficult to measure, but it had been estimated that for every 1,000 feet above 5,000 feet, work capacity will decline by three percent. So, by the time we injected oxygen into the system we may have suffered close to a thirty percent reduction in our work capacity. In fact, we had to scale down some of our operations in response. This was an additional factor in favor of adding oxygen to Biosphere 2.

For several months starting in fall of 1992 before the injection of the extra oxygen, Roy, Jane, Linda, and Taber received oxygen supplements through tubing that ran to their rooms. This oxygen was concentrated out of the Biosphere 2 atmosphere, it was not an addition of oxygen from the outside. This was done to relieve a symptom called Cheyne-Stokes respiration which occurred while sleeping (sleep apnea). Cheyne-Stokes is triggered when the body senses that it is lacking oxygen. When a sleeping person with this disorder takes a breath, it can end in a sudden gasp which wakes them up. Luckily after the injection of oxygen, their sleep returned to normal.

Roy alerted Mission Control in December 1992 that if he found nothing exceptionally interesting in the December blood analysis, he'll recommend an oxygen injection to bring the oxygen levels back up to between eighteen and nineteen percent. He also recommended that sixteen percent be the cut-off point in the future because under that level the stress factor markedly increased. Several months later, Roy received a letter from the US Navy requesting information on the experiment. The Navy was interested in applying this new knowledge to their submarine program—they wanted to maintain submarine crews at lower oxygen levels because this minimizes the danger of fire.

Once the full amount of oxygen was injected, all of our symptoms disappeared. Linda and Mark began to run joyfully around the perimeter of the domed west lung. Our friends on the outside watched us with wonder and curiosity as we all exuberantly laughed and talked with each other. Just minutes before we couldn't have climbed a single flight of stairs without panting at the top. What an experience! First, we endured a slow drop over sixteen months in oxygen with no corresponding change in air pressure. Then, suddenly, we received a large input of oxygen that gave us a huge lift. The eight of us will probably never take oxygen for granted again.

The wilderness encompasses an ocean, mangrove marsh, rainforest, savannah, and desert. Here marsh biome litterfall biomass is collected, one of many wilderness research projects.

CHAPTER 7

The Wild Side

"Good ecological engineering involves incremental changes to fit technological
operations to the self-organizing biota. The management process during 1991-1993
using data to develop theory, test it with simulation, and apply corrective actions was
in the best scientific tradition."

— **Professor Howard T. Odum**, University of Florida

THE CARE OF WILDERNESS

INSIDE BIOSPHERE 2, the wilderness cannot be taken for granted. On planet Earth
most non-tribal people except for explorers and naturalists are only vaguely aware
of the wilderness that sustains and nourishes life. Inside Biosphere 2, none of us can
escape realizing the importance of the wilderness areas to our lives and the majesty
of its operation.

SBV created the wilderness biomes from those which exist in the tropical
regions of the planet. Teams composed of ecologists, entomologists, engineers, soil
geologists, botanists, zoologists, microbiologists, and climatologists selected and
combined species, communities, soils, rocks, and water flows to make small but
richly diverse and characteristic ecological systems. These biomes required both
humans and a technical infrastructure to maintain the climate and certain other
conditions necessary for their survival. They were not only beautiful, but essential
for maintaining a breathable atmosphere and sustaining life. The wilderness biomes
also provided us and our animals with a small but important part of our food.
Their contributions to our well-being and the awesome beauty of our world were
incalculable. We were more than happy to reciprocate by maintaining the environmental conditions and ecotechnic designed systems necessary for their complexity
and diversity to flourish.

The biomes selected by SBV were the grand types of Biosphere 1: forest, grassland, desert, ocean, marsh, agriculture, and city. To maximize productivity of this
basic range of biomes, they were designed to be tropical. And from the range of
tropical types, the biomes our ecological designers drew from included regions
around the equator and some with monsoonal rainy climates. Thus our wilderness area comprised everything but the agriculture and city. Even the majority of

our staple agricultural crops were tropical. The wilderness areas were crossed only by small trails like tribal people use, or as a naturalist walks through the vegetation with minimum disturbance to off-trail areas. The wilderness areas were mainly managed by maintaining their required climate (temperature range, winds, rainfall, water chemistry, currents, and waves) and letting the systems organize and evolve. In cases of an aggressive or invasive species, the humans intervene as 'keystone predator' to help maintain biodiversity and ecosystem health.

Biosphere 2 was a new kind of laboratory in which ecological mesocosms of the vast tropical regions of planet were studied to understand how they functioned and what role they played in the global cycling of specific elements and compounds. We could arrive at deeper answers to such questions as: "What does a rainforest do for a biosphere?" or "Are all these different kinds of biomes necessary?" In addition, these tropical biomes are now among the most threatened ecosystems on Earth. Re-creating them in miniature and intensively studying them within a sealed system provided valuable information about how we might restore damaged natural areas. Thus, Biosphere 2 may have been the most important restoration ecology experiment ever undertaken.

Packing five different kinds of wilderness biomes in a space that was only about two hundred by five hundred feet constituted a unique challenge which attracted many of the outstanding wilderness experts in the world, including Dr. Richard Evans Schultes of Harvard, Sir Ghillean Prance, who was to become director of the Royal Botanical Garden at Kew, and Michael Balick of the New York Botanical Garden. The different biomes must be separated from each other by special transitional regions, called ecotones. Between savannah and desert was a thorn scrub; between rainforest and ocean was a bamboo grove; the mangrove marsh has both fresh and saltwater ecosystems; and between the marsh and ocean was a wall separating them. Each of these ecotones demands special conditions, intermediate between the biomes.

The ecotones might have been taken over by an expanding biome if we couldn't keep them at different temperatures and rainfalls. Air handlers and air vents, sprinklers and pumps made us masters of the climate inside Biosphere 2, with one exception: we could not control the sunlight. The tropical biomes we had replicated inside our biosphere were used to more intense sunlight, and we were eager to see how well they could adapt to the sunlight and day lengths in the latitude of the temperate region of Arizona. But the most critical question was whether or not we designed and built the biomes well enough that they themselves would tend to preserve their own diversity and integrity—with a little human help.

HUMANS AS NATURALISTS, HUMANS AS PREDATORS

We have to be able to sense if an ecosystem or species was in trouble before it became too late to do anything about it. Gaie had spent years of her life living between the marshlands and ocean, as well as exploring the world's oceans and diving among coastal reefs. Linda worked in deserts, Mark in deserts and savannahs, and all of us had made expeditions to rainforests. Our consultants were not only outstanding theoreticians, but at home in the wilderness biome they loved. These field experiences were essential because they alone could give a standard by which observers can assess the health and composition of our artificial ecosystems. We had a sense of how each biome should look, feel, and smell, and these observed sensations gave the initial clue if something's up—that something needs further observations to understand. SBV required each of the biospherians to have naturalist skills for living and working in Biosphere 2, and our stay there has helped hone our skills.

In addition to developing naturalist skills, the research teams selected some of the more complex and highly evolved species that functioned as 'bio-indicators' of the health of the overall system. These species were the first to show symptoms if their eco-region was deteriorating. Sometimes, these were the natural keystone predators, the species that kept other species of the food web in check, and whose continued existence showed if all the underlying links were still present. In the ocean for example, the corals and giant clams were great indicators of change; both quickly show the first signs of trouble when the system was stressed. Overall, the ocean biome itself had become a key indicator of the overall health of Biosphere 2 as we learned to watch its health by learning from the corals themselves.

We humans intentionally played an expanded version of the role of keystone predator for those parts of our small world which lack a natural one. We were on call to intervene when required to cull out predators that became too large or too numerous and so threatened the stability of ecosystems because of the decimation of their prey species. We intervened when a plant or animal species threatened to disrupt its eco-region or invade others. Learning when and how to intervene was a fascinating part of watching our biomes grow and mature. The mission rule was to keep interventions to a minimum.

Intelligent and judicious intervention by the biospherians assisted Biosphere 2's move toward self-regulation. Without human assistance, this new world would have lost many of its wilderness species. First, we depended upon machines which performed functions which were done naturally in Biosphere 1 to assure the health of the system: temperature regulation, recirculating water, providing rains, winds, and waves. Second, the wilderness ecosystems required individual and continuing attention. For example, the ocean and marsh required approximately four hours of attention and care a day. It was especially delightful to see how the systems immediately responded to our efforts—for example, when we installed the skimmers, a new system of water purification, nitrates quickly declined in the ocean and visibility improved.

Another major factor which the human contributes: the movement of biomass. In Earth's few true remaining wildernesses, this is done without human intervention: the massive power of organisms from microbiota to insects to animals consume, churn, tear apart, scatter, deposit, tunnel, displace, crush, and compost matter. Inside Biosphere 2, we couldn't wait for these processes, which evolved over millions of years on Earth, because many of them are too slow. We had to concentrate and accelerate the evolutionary process and help manage, as well as creating carbon sinks. Approximately 11 pounds or more of dried algae matter was removed from the ocean a week and stored. Tons of savannah grasses were collected, dried, and stored as well. We collected compostable material from the agriculture area at a rate of one hundred pounds a day. We fed the animals sixty pounds of fodder a day, and pruning the wilderness and savannah grass produced about one hundred and fifty pounds of material per week. On top of this, we produced about 270 pounds of human food per week.

This process helped maintain the elemental cycling which underlies the continuous operation of the Biosphere. It was a new role for people. In here, we altered or speed up ecological processes consciously to encourage environmentally beneficial results. For example, without herds of grazing animals like giraffes or deer, we humans had to harvest the savannah grasses before the rainy season begins, to encourage new growth. The edible parts are given to the livestock. The non-edible material was dried so as to prevent decomposition and respiration into our atmosphere in the form of carbon dioxide. At another time, generally in the longer day length months, when our atmospheric carbon dioxide levels permit, the cut grass was returned to enrich the savannah soils.

DECIDING ON THE WEATHER

In Biosphere 2, we could actually *do* something about the weather. In fact, rather than weather forecasts, we had weather requests. It was one of our most important jobs to keep our biomes happy about the weather. Humans programmed the computers to regulate the air and water temperatures, humidities, rainfall, and breezes while making sure pumps continued to make tides, waves, waterfalls, and the meandering flow of water in stream beds. And the observations and sensors gave feedback that enabled the system to improve itself.

Each of the biomes had substantially different requirements. The rainforest received moisture year-round, and temperatures stayed between 55° and 95°. The savannah was active for a portion of the year, but during its dormant season it received no rain. The thorn scrub and desert areas also received rain only part of the year. Natural coastal fog desert regions required hot dry summers and cool wet winters, when the little rain falls that they got, for optimal plant health.

The marsh was active year-round and relied on water flow and rainfall to maintain its salinity requirements for both its aquatic and marine life. Since it had freshwater marsh as well as salt-water marsh ecosystems, the sections of Biosphere 2's marsh were kept distinct from one another. The fresh marsh received daily rain, while the rest of the marsh had sporadic rain. Once rain entered the salty water, it was not a usable water source for the terrestrial biomic systems and had to stay in the marsh. We aimed at raining just enough to make up for whatever would evaporate, so Gaie frequently monitored water levels there. In the event of a possible flood (e.g. a rain-water pipe bursting, spewing water into the marsh), we used a machine that desalinates the marsh water and sends freshwater back into the rest of the wilderness water reservoirs.

Our ocean was tropical, emulating coral reefs and lagoon areas found in the Caribbean. Here it was crucial that water temperatures vary little between seasons, so our ocean maintained temperatures between 76° and 80°F. The Energy Center helped us keep seasonal conditions that were needed; it supplied hot water in the winter to heat the ocean and cold water in the summer to cool it. The beach, with its coconut trees and other salt-tolerant sand ridge plants, received rain several times a week, but aside from that, there were no rainstorms in the ocean. Instead, water was condensed out of the atmosphere, recycled through a rain tank, and put back into the ocean to replace what had been lost through evaporation.

The wind patterns provided the means to regulate temperatures in the different biomes. Even though there were no walls separating the rainforest from savannah or desert, the wind flowing from differently heated or cooled air handler units maintained the required air temperatures and humidities in these biomes. Winds also played an important role in pollination and air-borne ecological processes, not to mention the pleasure they provide to humans. But the maximum Biosphere 2 breeze was just five miles per hour—no squalls or hurricanes in there!

To create the climate for each biome, biospherians programmed control units which regulated the cooling or heating water that flowed into the thirty air handler units or into the heat exchange system for the ocean, both of which were located below ground. The Energy Center supplied three sources of water for this purpose. The coldest was the chilled water supply which was generated from ammonia freezers at about 40°- 48° F. Next was the water from the evaporative water cooling towers at approximately 50°- 65° F, and then the hot water which was supplied from water warmed by the waste heat from the electric generators in the Energy Center at 180° F.

The water entered Biosphere 2 in pipes which remained closed to the internal environment. These pipes ran through the Biosphere's walls carrying water that never came in contact with air or water sources inside. Once the energy was exchanged in the air handlers or the ocean heat exchangers, an appropriately hot or cool air or water would be blown or pumped back out into the biomes. In the summer this generated cool breezes in the tunnels and warm spots in the cold ocean.

Hot water was only needed in the wintertime to ensure that temperatures remain tropical and don't fall below minimum levels. By far our greatest energy needs were in the warmer months when we had to use cool and chilled water to prevent temperatures from rising past what life in our biomes could tolerate. Without cooling water in the summer, Biosphere 2 could rise to above 150 degrees in just a few hours. That's why we had backup generators and a second set of closed loop water piping between Biosphere 2 and the Energy Center in case the first set of lines sprang a leak.

We not only programmed our automatic devices on when and how long to rain, but we choose the quality of the water that was used. A continual challenge in our water recycling was managing water quality, the amount of nutrients and salts contained, pH, and dissolved solids. The rainfall in each biome produced water that drained through the soil. This water, salty and high in nutrients, is called soil leachate, and was collected in the basement and stored in tanks to be reused. Therefore, a process of testing the water commences in order to determine if it could be directly rained back onto the wilderness. Often the collected leachate water was mixed with condensate water (a very pure water condensed out of the atmosphere) in order to dilute the salt content.

The terrestrial biomes did not like salt. The ocean and the marsh, on the other hand, handled the salts but could not tolerate high levels of nutrients. We kept a careful eye out for the build-up of either salts or nutrients with the long-term goal of assisting the cycling of certain elements in our biomes. Occasionally the agriculture system would also be involved if it ran out of condensate generated from the IAB and needed condensate water from the wilderness. At times, we had intense negotiations over allocation of condensate water, each area manager arguing his or her case for receiving more.

The 'rain' came mostly from sprinkler heads mounted overhead in the space frame. In some special spots (where we didn't want to get walkways or ventilation gratings wet), there were also ground sprinklers with directional heads. As trees grew in the rainforest and started to interfere with an even flow of falling rain, we laid down soaker hoses to ensure that every area received adequate moisture. It was extremely important to provide the right amount of rainwater.

Some of the most delightful contrasts came from watching our biomes go through their seasonal changes, as outside, the high desert mountain ecology of southern Arizona goes through a different series of seasonal changes. It was often quite strange: walking through a light rain shower in the rainforest while outside it may be winter and snow covered or sizzling at 105 degrees under the dry desert sun. Conversely, we may be working in the dry and dormant Biosphere 2 desert, while the sky is black with rain clouds and the landscape outside is inundated with a monsoonal downpour. At such times, with the raindrops beating on our glass roof, we could almost taste the outside rain.

In some areas where slopes were steep, like in the east ginger belt of the rainforest and the desert bajada, we were careful about soil erosion. After closure, we spent many hours replanting ground-cover vines and plants in the rainforest where the footsteps of construction crews and guest tours before closure prevented them from becoming established.

Even with such 'bonsai' biomes, we were very careful to avoid excessive trampling. Narrow trails were established in each biome for biospherians to follow and we carefully tread as to not veer off the track unless management or research required it. Same with the coral reef, all of us enjoyed snorkeling from time to time in the ocean but all knew to keep their feet and hands off the corals.

One of the ecosystems in the rainforest required year round fog. This was the cloud forest area of the rainforest modeled after the Amazon cloud forest 'tepuis' where plants are naturally covered in almost permanent low-lying clouds. To produce this fog, we used artificial misters. Every five minutes, spray nozzles ejected a fine mist around the top of the rainforest mountain bowl in which are planted ferns and vines around several small ponds. When the misters were on, visibility could fall to just six to twelve inches—a man-made cloud.

There were two streams in Biosphere 2: one which meandered in wide loops through the varzea (flood plain) of the rainforest, and one bordering the African acacia trees to create the shady 'gallery forest' of the upper savannah. Biospherians regulated the flow in these streams, periodically cleaning out any vegetation that may choke the passageway and ensuring that their recirculating systems were operating. The varzea stream water was continually pumped back up to the interconnected ponds at the top of the cloud forest. From there, water overflowed to create the waterfall that tumbled some twenty-five feet to the pleasantly cold and deep 'Tiger Pond' at its base and thence back into the varzea stream. The savannah stream, when it reached its terminus at the southern border of the upper savannah, was automatically pumped back to its head, where a small waterfall cascades down to commence the stream run.

HELPING THE RAINFOREST GROW

The rainforest was one of the most challenging terrestrial ecosystems we had to build from scratch. Arizona is far from the tropics where day lengths vary only slightly from season to season, and to start with, the rainforest design had to protect the light-sensitive plants from the blazing desert sun overhead. Dr. Ghillean Prance, then Vice-President of Research at New York Botanical Garden, and Director of the renowned Royal Botanical Gardens at Kew near London, was one of the world's greatest experts on the Amazon rainforest. He came up with a strategy that he thought would not only be successful for Biosphere 2, but also be a first trial for a useful method for reversing the devastation of the Amazon itself. The first element was a fast-growing selection of trees that quickly formed a canopy to shade and protect the slower growing rainforest trees that constituted our mature system. While those first canopy trees formed a protective shield against overhead sun, a thick gingerbelt planted around the periphery of the rainforest screens out the harsh sunlight coming in on the sides. The gingerbelt was populated by true rainforest species, but ones which tolerate strong sunlight and are fast growing: banana, canna lilies, heliconias (bird of paradise flowers), and ginger.

When the biospherians worked with a team from the Yale School of Forestry in 1990 and 1991 to measure the initial size of all the plants before closure, the tallest trees were under fifteen feet. It required considerable imagination to picture this scrubby scene as the site of a future noble rainforest. The bamboos of the bamboo belt, which were planted along the beach cliff to protect the rainforest from salt particles blown from the ocean, were just a couple of feet tall. The varzea was still exposed to harsh overhead sun and the soil along the stream banks was held in place by burlap as a stop-gap measure until the plants sent out their roots to hold the soil in place.

In any event, the rainforest thrived during the first two years of our closure experiment. By the first anniversary, biospherians logged many hours cutting back trees that were touching the space frames fifty feet above the ground. Some of our first-canopy trees, Leuceana, Cecropia, and Ceiba, grew more than twenty feet the first year. Almost overnight, they created a rainforest with towering trees protectively shading the lowland rainforest. The bamboo belt had clumps that were over fifteen feet tall (now, they are approaching thirty feet). In the varzea, modeled on a 'black river' area of the Amazon, the small Phytolacca trees we planted grew both up and out. Three trees now clothed the varzea with heavy shade, with tree trunk diameters that one could barely stretch your arms around.

The gingerbelt also worked as designed. It became so dense that it completely blocked out views of the Mission Control building a few hundred feet to the east. Because it's fast to regrow, it played a valuable role in our management of wintertime carbon dioxide. We pruned the gingerbelt back and stored the biomass dry in our technical basements to prevent decomposition and release of carbon dioxide.

Meanwhile, since the gingerbelt rapidly regrows, the plants very effectively take up carbon dioxide. The gingerbelt could be cut two or three times during the fall and winter months if required, and in the summertime, this dried biomass was returned to the soil to release needed nutrients.

The rapid growth of the first canopy trees created a problem of its own. Because the trees grew so fast and because there was less overall light inside Biosphere 2 (due to light reduction caused by the glass and space frame) as compared to the natural Amazonian levels, their limbs became weak as they stretched for light. Their top growth became disproportionate to their root growth. That, combined with the lack of strong winds, reduced the necessity for trees to grow deep roots and produce 'stress wood'—the stronger wood that allows trees to handle strong wind impacts. So, some of the tall rainforest trees started to lean and a few had toppled over. In the upper savannah, several acacia trees, which had also grown tremendously, arched right over or had their top branches break.

Because of this, one of the periodic wilderness jobs was to decide whether to straighten the trees with braces or take weight off them by pruning back some of their top branches, thus allowing the roots to catch up in better anchoring them. The former decision calls for biospherian acrobatics. One of us climbed high in the space frame to attach a rope from the roof to the tree and then it is winched more upright. In some cases, we left the rope as a long-term assist; in other cases long wooden braces are cut and attached on three sides of the tree to prevent any further leaning.

TALL AND WEEDY TALES

One of the dire forecasts that some scientists made before we began the experiment was that we wouldn't be able to maintain separate biomes because weeds from one would invade others; and each would be swamped by its own rankest growers. While that hadn't proved true, a few plants required considerable effort to control.

During the first six months of our stay in Biosphere 2, we all admired the morning glory on our weekend strolls through the biomes. Covered with blue flowers, the morning glories draped twenty feet down the sides of the cloud forest. The modest growth of the domesticated flowers you may know from your garden gives little indication of what prodigious feats a morning glory is capable of given an environment it loves. Our jungle morning glories were not content to cover the sides of the cloud forest. Soon they spread a dense cover of their leaves over trees on every side of the cloud forest. They even forded the Tiger Pond (which stands at the base of the waterfall), clogging the water with hundreds of pounds of vines and roots. Then they marched south along the ground and into the trees of the varzea and clear over towards the beach as they fell over the cliff. Wherever their long vines dropped on damp soil, they put out roots.

Morning glory probably started its life by the planting of a handful of seeds on the sides of the cloud forest mountain where it was expected to remain. It must have produced several tons of biomass during its reign as the King Kong from the cloud forest. At first, we welcomed the fast growth and its shade, also thinking of all the carbon dioxide the vines were soaking up. In addition, periodically cutting them back provided delicious fodder for the goats. But as time went on, our strategy of fast-growing trees succeeded well in providing shade for the slower-growing trees. The dense covering of morning glory leaves was too much. Tree branches sagged under the weight of the invaders, and understory light levels fell disastrously, even for rainforest species. As keystone predators, we decided that morning glory was not going to be part of our rainforest since it takes far too much time to keep under control. It had to be weeded out.

By the first summer we started waging war against the morning glories. But we'd let them go for too long. Their vines were everywhere on the forest floor—a thick maze which tapped into the soil for hundreds of square feet from where they climbed down from a nearby escarpment of the cloud forest. It was no longer possible to tell where any of it started or ended. And any bit of the vine left uncut or uncollected regenerated itself and relentlessly grew again. When we halted the morning glory advance they were about to cascade from the varzea and start the invasion of the upper savannah to the south!

In the final year we followed this with a systematic campaign to find and eliminate every last bit of morning glory vine before the end of the two-year mission, so that it wouldn't be a burden on the next crew. Taber and Linda went into the space frames to liberate the trees from the vines (working before breakfast before the temperatures up there became too hot). Mark did much of the ground patrol work, cutting and bundling up morning glory vine like so many rolls of electric line.

There had only been a few other plants which, like morning glory, proved to be serious weeds or required much time for cutting back. One of them, curiously enough, was the passion fruit vine. Passion fruit was a planned member of a couple of our biomes. In the rainforest, we planted several varieties of this tropical plant to provide genetic diversity and food. In the savannah, it was planted under some of our African acacia trees in the gallery forest that fringes the savannah stream. We were rewarded with a rain of passion fruit during the summer and fall months.

Passion fruit makes a tasty ice cream flavoring and lent a special zing to our fruit juice mixes for feast drinks. But, somewhat like the morning glory, the passion fruit vine demonstrates how prodigious plant growth can be given the right conditions. A vine on the rampage invaded new territory and smothered whatever was in its path. Our savannah passion vines covered whole areas of the savannah in a grape arbor-like web overhead, bending tree branches with their weight. In the east rainforest, they had ascended from the top of the gingerbelt plants and created an impenetrable mass of leaves which covered the higher space frame glass.

In the summer, we let the vines grow when there was abundant light for the other plants. This gave extra shade to sensitive trees underneath. In winter, when light was very limited and our rainforest trees went half-dormant, we pruned back the passion vine to let more light reach the under story plants.

In the desert and lower thorn scrub, the major invader was Bermuda grass. We suspect it came in with some of the local soils, but with the year-round tropical climate of Biosphere 2, it was far more productive than in the deserts of Arizona. Compared to the morning glory, however, the grass problem was far less serious. For one, the desert and thorn scrub areas went dormant for part of the year and, without water, the Bermuda grass couldn't grow. It had been more of a problem in the sand dune area of the desert. There were more empty spaces there, and a variety of grasses grew in between the sand dune bushes and the creeping devil cactus.

The marsh required weeding of the coastal grassland and especially the cattails, which were abundant and like other invasive species, are dominant. But in the rest of the marsh ecosystems, the mangrove trees required little extra care. They too were rapidly growing towards the light while taking up carbon dioxide.

By our second year, the ocean also started to have its share of weeds. Macro algae is nature's best remedy for removing nutrients and taking up CO_2, and ours were doing a great job, except they too were threatening to outcompete the coral colonies. So Gaie dived four or more times a week to pick off the colorful, large algae (and especially the green bubble algae) from around the perimeter of corals and from the ocean floor where it accumulated in floating islands. In addition, Gaie routinely removed a 'grey slime' that appeared after closure and started growing all along the limestone reef surface, slowly encroaching its way along the edges of the coral colonies. This expanding goo is a combination of green algae, cyanobacteria and sediment, as well as a home to small crustaceans called tenaids. Freshwater chlorella algae are notorious for being one of the fastest growing algae species; the Russians had used it for air and water regeneration in small closed systems. So it's not so surprising that a marine species of chlorella found its way into the system and took advantage of the high carbon dioxide and available nutrients to grow everywhere. As with all the wilderness biomes, the coral reef also had its share of weeds and the biomass is dried and stored, adding to our carbon sink.

Part of the maturing process of our ecosystems was their resistance to the invasion of foreign weeds. We probably did more weeding than future crews would have to do, because the biomes had not yet evolved a hardy, ecological community. We had already begun to see in these two years that, as trees get taller and their foliage shades the ground more, it became more difficult for weed invaders to grow below them. In the upper thorn scrub, the grass problem was already far less serious than it was at the beginning of the closure since the thorn scrub trees doubled in height. In this sense, we biospherian keystone predators had lent a helping hand in these early years while the biomes organized themselves.

THE DESERT SHOWS A MIND OF ITS OWN

Life asserts itself in surprising ways, and this is particularly true with the desert. When SBV designed it, a basic problem had to be solved: how to find a desert that could live with the high humidity which could not be avoided with an ocean, marsh, and rainforest under the same roof. The research team found the solution in the 'fog deserts' of coastal regions. For part of the year, heavy fogs rise from the ocean currents and cover the vegetation, supplying some of its water needs and reducing transpiration losses. These deserts have adapted to high humidity, though rain is still extremely sparse and sporadic as in all desert regions.

Fairly close to Arizona is one of the most striking fog deserts—that of Baja California. Collections from the plant life there formed the majority of the Biosphere 2 desert, although some plants from Israeli, African, and South American coastal deserts were also included. The semi-scrub plants, desert bushes, and annuals (whose seeds were collected from Baja) grew luxuriantly in our conditions, enjoying the longer rainy season we gave them the first year in our attempts to keep carbon dioxide levels low. In places where condensation off the glass added extra moisture and extended the growing season before the soil dried out, they did especially well. The salt bush family (Atriplex) formed a bonsai forest on the slope down to the salt playa. It was only in limited areas of the desert that the Boojums, yuccas, and columnar cardon cactus (a fog desert relative of the Saguaro cactus which grows on the mountain slopes outside Biosphere 2) dominated, giving the kind of cactus and succulent desert originally pictured.

The bulk of the desert developed into what ecologists call a coastal sage scrub desert environment. Instead of trying to force it to stay in the original conception, we went along with its own tendencies. We included a good diversity of plants in our original introductions so there was still a good mix of species. We would introduce yet more diversity during the transition, enriching the kind of ecosystem that was emerging on its own.

The two thorn scrub ecosystems had developed as anticipated. They were put in as transition zones (ecotones) between the savannah and the desert, for that is how they are frequently found in the tropics. The upper thorn scrub was primarily based on communities found in Sonora, Mexico, and the lower thorn scrub is dominated by Madagascar/South African thorn scrub species. The night-blooming cereus cactus from South America was a special plant in the upper thorn scrub. Its beautiful white blossoms appeared for just one or two nights at the beginning of the rainy season.

In both desert and thorn scrub, the biospherian weather-makers had to pay careful attention to the plants to gauge when to start the rains and when to end them. This was far more important than in the savannah, which has adapted to a variety of rainfall seasons and more frequent out-of-season storms. The desert and thorn scrub species would not withstand rainy seasons much different from those of their natural habitats. Since many vectors are somewhat different inside Biosphere 2 (such as day lengths, temperatures, the mix of plants) we watched the desert and thorn scrub carefully to pick up signs that the plants are ready for rain. Certain indicator plants are frequently checked to see if their leaf buds were swelling or if flowers had appeared on those which bloom just at the beginning of the monsoon season in their native habitats. When the plants signaled yes, the irrigation controls which program the rain were set. The word was passed on: "On Thursday at 5 PM, the north desert will receive its first rain;" or "Saturday morning right before breakfast, the upper thorn scrub will get an initial rain lasting forty-five minutes."

We also made frequent checks of the desert and thorn scrub to gauge when the rains should be stopped. While the desert plants could withstand a short rainy season or a drought year, they may have a harder time with too much rain. In these initial years of operation, we wanted to keep them active as long as possible, so we checked key species for signs of root or stem rot that might indicate we were keeping them active too long. When the rains are stopped, the biome dries out and becomes inactive again.

HOME ON THE RANGE, WHERE THE TORTOISES GRAZE

Our savannah was more flexible than the desert. It contains a wide spectrum of grasses from all the world's savannahs—Africa, Australia, and South America. The challenge in this biome was to replicate the feel of the open horizons of a savannah in barely half an acre. The savannah was long and narrow, bounded by the cliffs that separated it from the ocean and marsh to the east. The whole structure sloped to the south, where the desert lies at the lowest elevation in Biosphere 2. This allowed hot, moist air to rise and flow uphill towards the rainforest at the north end of the wilderness biomes. The savannah was also divided into an upper savannah where the stream and gallery forest of acacias and passion fruit are planted, and the lower savannah which formed a small ocean of grasses.

Savannahs are tough environments, exposed to seasonal flooding and long dry seasons, with far more scorching temperatures than the rainforests. The grasses which dominated them are a recent evolutionary addition, and they grow continuously to survive the teeth of grazing animals. In fact, most savannah ecologists agree that savannahs need disturbance to stay healthy. Grazing stimulates the grasses to extend themselves with underground rhizomes, and to put up fresh green growth. Most grasses have adapted to the heavy hooves of range animals and to periodic fires which clear out the old growth. But fires were forbidden in Biosphere 2 (because it would pollute the atmosphere) and our savannah was far too small to even think about introducing any of the large grazing animals which form the herds of the African savannahs.

At one point, we contemplated the idea of occasionally taking our goats out on leashes to graze. This was impractical for logistical reasons, so we looked for wild grazers that could be supported. We chose leopard tortoises from Africa which were introduced as a herd of three before closure. But we had few illusions that these twelve-inch-long animals could make much of an impact on our grasses!

Put a sickle in your hand, and start to hack your way through the savannah, and its size suddenly becomes magically magnified. We experienced that the first fall. The lower savannah, grown into a crowded tangle of four-to-eight-foot high grasses, was cut step by step by advancing armies of biospherians. At the end of several weeks, our little plot seemed like vast grasslands. When we shifted to the upper savannah, the billabongs also commanded respect. These areas were named for the low-lying semi-lakes of Australia, which are actually running waters during the wet season, but which gradually shrink in depth and sometimes dry out completely during the off-season. The Biosphere 2 billabongs were about twenty-five feet long and twelve feet wide, and seasonally fill with the overflow from the savannah stream. But try to cut your way through them and you discover what kind of growth some grasses are capable of! We pulled out some that were twenty-five feet long.

The African acacia trees had also made themselves at home. When they were planted, just a year before closure, they were thin and straggly. Now nearing re-entry, several of these trees were approaching the roof, thirty-five feet overhead. Festooned with passion fruit vine and other vines from the savannah stream that hitched a ride on their branches, they were beginning to give the landscape the special look that acacias give savannahs worldwide. While the grasses learned to live with the grazers by outgrowing their teeth, these African acacias employed a different strategy: nasty thorns to convince the wildebeest and giraffe that acacia leaves are no free lunch. We traveled the savannah in thorn-proof shoes, pruning shears at the ready to lop off dangling branches. In this respect, the acacias were like the thorn scrub areas which divide savannah from desert.

THE WATER WORLDS

The 72,000 gallon marsh system was modeled on the Florida Everglades. It was designed as an estuary with six different ecosystems, where water flowed from the coastal prairie to white, black, and finally red mangroves. There were 11 pumps for each of the six different ecosystems to provide currents within each particular water body and 12 algae scrubbers were designated as for its nutrient removal system. The design followed Dr. Walter Adey's (Smithsonian Institution) mesocosm approach, which he had developed in the eighties with SBV.

The first prototype system was built in the basement of the Smithsonian and proved successful, thus we replicated that approach for Biosphere 2. This entailed removing bulk sections of the Everglades marsh, including sediment, water, plants and associated fauna and flora, into 1.8 cubic meter plywood sealed boxes. These boxes were then trucked to Arizona and held in a temporary greenhouse that was outfitted with all the technological support systems intact. This included a tidal system, currents and a nutrient removal system. It was a massive undertaking that took the helping hands of many, but particularly of Gaie and Laser, as well as Malone Hodges, marsh biome assistant, to support the little marsh on its long journey into Biosphere 2.

The biggest challenge, oddly enough, was obtaining the permits for the marsh to cross from state to state and in particular when the first truck reached the Arizona border, it was stopped for days. This was a critical moment as the marsh boxes had to be hooked up as soon as possible to their tidal life support system awaiting them at Biosphere 2. After many phone calls and meetings, the problem was solved: with great relief by all, the inspectors finally understood that these trees were mangrove trees, and not mangos!

The ocean biome was modeled after a Caribbean fringing reef with a beach and lagoon that had a channel leading to the open ocean around a reef crest and fore reef that descended into a deep sandy shelf. Because Biosphere 2 was hundreds of miles from the nearest coast, we had to recreate the reef using local materials, such as calcite limestone rock and sands to build the reef structure. However, it was with necessity that we also transported from afar aragonite shells from Texas, and coral sand and algae reef rock from the Bahamas to aid in proper ocean chemistry, some ocean water from the Pacific Ocean with critical microfauna and flora, and corals and fish from the Yucatan Peninsula. The most challenging was the collection and transportation of corals. Gaie and her expert team of marine divers (Laser and Taber) went to Akumal, Mexico to remove entire coral colonies with a calcium carbonate base that could be mounted back onto the Biosphere 2 reef using its reef structure and/or epoxy. These accessions, with associated fish and invertebrates, were transported overland on a 3.5 day journey in trucks that were outfitted as coral aquaria systems including lights and currents, and the help of Mexican officials to ensure their safe arrival!

Construction of the 900,000 gallon ocean began in 1988 with a welded steel tank that had a maximum depth of 26 feet with three ocean windows: one in the lagoon, one at the fore-reef, and one in the deep ocean for viewing from the outside. The water was introduced in 1990 and recirculated with specially designed vacuum pumps and wave machine, before the other accessions followed. Of this water, 180,000 gallons was transported (30 loads) via dairy farm trucks from the shore off Scripps Institution of Oceanography and the rest was made on site with a sea salt mix called Instant Ocean. A few months before closure, however, and much to our dismay, we found that our 20 vacuum driven pumps were failing frequently and could not sustain the water flows required. Thus, they were quickly replaced with standard seawater pumps.

Marshes are known for their high levels of organic sediments, high-nutrient water sources, and richly diverse marine plant communities. Coral reefs are entirely the opposite, with porous, sandy sediments, low nutrients, plenty of algae, but only a few underwater plant species. Key species in a tropical marsh are the mangroves: white, black, and red mangrove trees reaching into the salty water, masters of bridging the terrestrial and salt water realms. The peculiar branching root structure of the red mangroves, gray-white oyster beds that rise out of the water, and dense green foliage hanging suspended over dark green and brown marsh water are the characteristic traits of the mangrove marsh ecosystem, forming a network of entangled prop roots. The coral reef, on the other hand, is the secret realm of the ocean. If the observer gazes on top of the water, there is nothing to see except coral heads here and there breaking the surface of the water. But under the sea is the colorful, dazzling, and magical reef world with caverns, currents, fish schools, and large sea animals that cruise in and disappear as quickly as they appear.

Coral reefs are notoriously difficult to maintain and the Biosphere 2 ocean was a big leap from the little that is known about aquarium exhibits. Both of these equatorial systems are now nestled in the foothills of Arizona desert mountains, approximately thirty-three degrees north of the equator, 3,900 feet above sea level, thousands of miles from the nearest tropical ocean, and controlled by the regimes of a temperate light pattern changing markedly in intensity and duration between seasons. The entire system was dependent on supportive technology. For example, waves were generated from an approximately fifty-foot-long vacuum system that sucks approximately ten thousand gallons of water into a ten-foot-high chamber and then releases it at the south end of the ocean to create a one-foot-high wave. With each suck and then release of water, it sounded to us like the breath of a whale. Metaphorically, the wave machine brought Biosphere 2 to life, and in actual fact, to the ocean.

The biospherians' role was to maintain the natural conditions and keep vigilant attention on both the ecology and technology. Monitoring the water chemistry of marsh and ocean gave us essential information about the health of the systems. Water samples are collected weekly for nutrient analysis, pH, dissolved oxygen, salinity and other parameters, and daily checks of all the technical systems are made to be sure they are operating correctly. The mechanical checks were critical because the water systems required motion and flows to support higher life forms. If water flows stopped for more than twelve hours, the system went on a red alert. Several times the wave machine caused problems, mechanical as well as computer controls, and by necessity Laser was on call to work as long as necessary to get the systems back up and running.

Initially, we operated the marine systems with a diurnal (12.5 hour) tidal system between the marsh and ocean. When the marsh was at low tide, the ocean was at high tide, and vice versa. This required pumps, controls and levers to move approximately 6,000 gallons of ocean water into the red mangrove marsh which flowed slowly up to the freshwater marsh section which is at a higher elevation. Once a high tide was reached, the water then flows back to the ocean controlled by a computer until the low tide level was reached. While the tide was beneficial to the marsh, the dissolved organic matter (tannins and nutrients) that returned back to the coral reef from the marsh could not be managed. There was not enough 'ocean water' between the two biomes to support the tides. Thus, in November of 1991, we shut the system down and from then on operated the ocean and marsh as separate water systems.

Once constructed, monitoring of the chemical and physical properties of the marine systems, as well as tending to its nutrient removal systems, continued to be our biggest challenge and defined the day-to-day actions of Gaie and her team members on the inside and outside. Because of all the unknowns and challenges, the marine team included several key individuals as advisors. Our marine chemistry experts were Dr. Robert Howarth (Atkinson Professor of Ecology and Environmental Biology, Cornell University), Dr. Roxanne Marino (Analytical Chemist, Cornell University) with whom Gaie consulted regularly.

One of our primary concerns for the ocean was whether its biodiversity and complexity could survive the dramatic fluxing of carbon dioxide. As the concentration of CO_2 increased in the atmosphere, it diffused into the ocean water and this lowered the pH of the water (making it more acidic). Coral reefs are found in waters where pH ranges from about 8.2-8.4. With the high levels of carbon dioxide that were predicted in Biosphere 2 after nine-week-long pre-closure experiments, we realized that we needed a method to 'buffer' the ocean in order to keep the CO_2 from drastically lowering its pH. Therefore, we imported the aragonite reef rock, shells and sands to augment the capacity of the system to withstand the change in pH and we stockpiled fifty-pound bags of carbonates and bicarbonates of which approximately 4,500 pounds were added throughout the two years to counteract the rises of carbon dioxide. This chemical effectively ties up the CO_2 as it dissolves into the water and increases the ability of the water to withstand an influx of CO_2. Even though this helped keep levels of pH from plummeting, the lowest level was 7.6 which was well below the pH of natural coral reefs. Curiously, 7.6 is known to kill small reef aquaria, but because the ocean was large and biodiverse as compared to any other aquaria system, it was able to withstand that drop in pH. We had no prior information about what happens should pH reach the levels it did. The ocean proved after two years to be far more adaptable than we had thought possible.

Most of the life forms in the ocean were the microfauna and flora; the unseen critical components providing necessary functions and adaptive responses. Thus, we hired microbiologist Dr. Donald Spoon (Assistant Professor, Georgetown University), to make an assessment of the aquatic microbiota of Biosphere 2 including the fresh, brackish and saline water basins. There were close to 2000 microbiota species, thus affirming we had the worker-bees for the wilderness water systems that would be critical to their healthy functioning.

For the bigger life forms, the biospherians became keystone predators of the ocean. It was a vast ocean with sea squirts, mollusks, crabs/lobsters, sponges, algae, a Thalassia seagrass, anemones, soft corals, hard corals, echinoderms (brittle stars, urchins, sea cucumbers, starfish), worms, protozoa and microbiota.

By year 2, the coral reef became plagued by an outbreak of fire worms, a red-orange, prickly worm that eats soft corals and, in particular, anemones. They were not intentionally introduced into the system, but slipped in with the reef rock transported from the Yucatan, Mexico where the corals were collected. They quickly grew into a serious problem. For several months shortly after closure, it was not uncommon to hear tour guides on the radio reporting a fire worm sighting through the ocean window to alert Gaie so she could quickly don her diving gear and remove the worm. Several hundred were removed, and although there were still some small ones found from time to time, close to closure, they were no longer a problem.

Another species that required culling were Spanish and spiny lobsters. Three lobsters had been introduced into the system as small adults and by the end of the first year, they had grown much bigger and began decimating the snail population. Snails are key grazers of small algae, and without them, the algae would smother our corals. After some lengthy discussions about the matter, the research team decided to remove two of the lobsters in order to give the snail population a chance. Additionally, all the queen trigger fish (excepting one that managed to evade capture), were removed in 1992, when it was realized they were consuming the giant clams. The parrotfish and squirrelfish also continued to be a problem. Individuals in both species had grown considerably. The parrotfish, the largest fish in our ocean (some grew to about a foot and a half long) from time to time took bites out of the corals. The squirrelfish patrols for new fish babies, consuming them before they can grow to adulthood. Laser tried to hunt the largest of the offenders but to no avail because the reef was quite porous and provided ample opportunity for the fish to hide. Techniques used commonly on Earth to catch fish in open water, such as harpoons and nets, don't work in this coral garden designed with complex econiches to hide in. This was undoubtedly an area where fine-tuning in management skills or changing the choice of species to complete food webs was required.

The marsh was the easiest of all our biomes to maintain. In general, the marsh had flourished. The mangroves were soaring twenty feet above the water; when they were introduced three years earlier, they were young trees with a maximum height of about five feet. Aside from the increased biomass, the marsh proved to be an essential component in closed systems because it was extremely resilient. It had the capacity to receive excess salts or nutrients in its sediments without causing damage to its life forms, grow rapidly, thus becoming a great carbon sink, and a beautiful addition to the wilderness area that was often unnoticed as one of the largest biome areas of the planet.

PASSING ON THE TORCH

Our biomes have come a long way towards growing up in the two years that we had been their allies and occasional caretakers. But they had a long life ahead of them, probably decades of continued growth to maturity. Though our biomes were modeled on the tropical regions of Earth, they became unique and distinctive ecosystems, adapting to environmental conditions unlike those found outside Biosphere 2.

*Hundreds of pipes, pumps, and other mechanical devices help maintain
the ecological environments inside Biosphere 2.*

CHAPTER 8
The Technosphere

"To live as human beings means to live with other life. The nauseating fear that machine technology would replace all living species has subsided in my mind. We'll keep other species, I believe, because as Biosphere 2 helps prove, life is technology. Life is the ultimate technology."

– **Kevin Kelly**, *Out of Control: The New Biology of Machines, Social Systems and the Economic World*

BIOSPHERE 2 HAD TWO FACES, one above ground and one below. Above ground were the abundant, varied life forms of the wilderness areas and farm, while the underground environment was like a submarine full of humming and whirring technology that supported the greenery above. In the basements were some one hundred twenty pumps, forty-five air handlers, several miles of electrical wire and pipes, water storage tanks, wastewater tanks, aquaculture tanks, computer controllers, filters, an algae-based nutrient removal system for the ocean and marsh, systems for rainfall irrigation, heating and cooling exchangers, desalination system, a chemical recycler for atmospheric carbon dioxide, diving equipment, composting equipment, agriculture processing machines, ovens for drying, and other assorted machines. In addition to this, a large workshop designed by Laser with mechanical, plumbing, electrical/electronics, welding, woodwork, and rigging equipment as well as spare parts which he crafted in order to handle any maintenance or repairs without having to rely on importing equipment and supplies from the outside. A complex network of computers, sensors, videos, and communications gear runs throughout Biosphere 2 which gave us real-time status reports on the technical systems. In addition, a computer-accessed system of vibrational analysis tracks many of the key technical components of Biosphere 2's systems and gave us advance warning of impending breakdowns.

The smooth operations of the technical systems had been an extraordinary success. This was a tribute to the care and workmanship that went into Biosphere 2's design and construction. Laser, manager of the technical systems, was able to work with the other biospherians to accomplish all maintenance and repair operations using less crew time than we anticipated. Only eight percent of all biospherian work time was required for the repair, maintenance, and upgrade of technical systems, which was less than thirty hours per week. In addition, Laser upgraded some technical systems making them easier to operate. He also fabricated entirely new systems that we didn't realize we needed until after closure.

During the seven years it took to build Biosphere 2, Laser served as head of Quality Control. He was involved in every phase of the development of our major systems from blueprints to installation. He knew he was going to be the one that maintains all of these systems during the first two-year mission. This ensured an active dialogue between construction and maintenance personnel during the critical design period. Usually maintenance men are specialists. They don't get involved with a system until after it is complete and then they discover the problems of actually operating it. In the case of Biosphere 2, Laser helped ensure that ease of maintenance was an integral part of the design. This fact is significant and may have a great deal to do with why our technical systems performed so well. One of the key design criteria was the choice of lubricants, oils, and greases. We could not afford to use toxic chemicals that are not recyclable, so all these substances had to be what is called 'food grade.' After considerable research SBV found what we needed.

One of the main ideas behind the maintenance system for Mission One was to simulate, as nearly as possible, a two-year voyage to Mars in which the crew relied on what was included inside the system. We carefully planned for everything we might need. Back-up pumps and spare parts were included as well as the tools and machinery, to fabricate virtually any piece of equipment. Laser was key to this strategy since he had most of the skills required for such a daunting task, and SBV sent him to intensive schools to acquire the rest.

For the first year, we operated with nothing coming in from the outside. In the second year our Scientific Advisory Committee recommended a limited amount of export and import to enhance the research program. We then allowed the import of spare parts if they saved substantial biospherian labor and the export of malfunctioning equipment such as electronic or computer components needed for our scientific research program. The original idea of total self-sufficiency was very useful, since it encouraged the biospherians to trouble-shoot their own systems and not rely in any way on outside specialists to fix something that was broken. This is where hands-on experience is invaluable. The person who 'can do' is vital to the operation of a closed system, rather than the textbook theorist who has never gotten his hands dirty. The ability to improvise and to 'jury-rig' equipment when necessary was a crucial aspect of living in a closed system.

MAKING THE BIOSPHERE AIRTIGHT

What greatly increased the significance and scientific research of the experiment depends on the virtual airtightness of Biosphere 2; that's what makes it a materially-closed facility. From the beginning of the project we'd identified this as the number one technical engineering challenge. It was the task of Bill Dempster, our head of Systems Engineering, to reduce to a minimum the rate at which air leaked out of or into the Biosphere, a task he worked at right up to closure and all through the first experiment.

Bill was notorious in the pre-closure months for spending long and tireless hours in the underground tunnels that surround Biosphere 2. He wore headphones to magnify the hiss of air caused by tiny leaks along the stainless steel seams and joints which he intended to seal before closure, and he carried a telephone to transmit the leak's exact location to his assistants. It was mind boggling that he had the patience to do this minute detection work on a structure that was so enormous!

Bill wasn't the only one looking for leaks. There were also 'space frame men' with soap-bubble machines. Over twenty miles of glass seals were covered with soap bubbles (just like you use to find holes in automobile tires) as the first step in tracking down air leaks. Over thirty leaks were discovered in this way. To use the soap bubble method, the Biosphere had to have a slightly higher air pressure than the outside. This forced air out of the holes and created the soap bubbles. Bill figured out that the best way to create such a pressure was by using two 'lungs' which he designed for Biosphere 2.

The lungs, which were housed under beautiful white geodesic dome coverings which protected the interior from harsh sunlight, are huge rubber membranes which were connected to Biosphere 2 by underground tunnels. As the air temperature rose during the day, the air inside the sealed structure expanded, rushing through the tunnels where the lung membranes rose to accommodate the expanding volume; at night, as temperatures fell in Biosphere 2, the atmosphere contracts and the lung membranes deflated.

This lung design enabled us to manage the air pressure and build the vast roofs of space frame and glass. Without the lungs, we would have had to reinforce our structure, or it might have faced the danger of exploding or imploding! By using the lungs to exert a slightly higher pressure inside the Biosphere than outside, we were able to calculate the amount of air leaving the system by tracking the height of the lungs and comparing that with our calculation of what the height would have been if there were no leaks.

In December 1991 we ceased creating a positive pressure to force air out in order to find the holes. Bill then relied on the concentration of two trace gases to show how much air was getting out. He introduced a known quantity of two biologically inactive gases, helium and sulfur hexafluoride (SF_6). In addition, the researchers from Columbia's Lamont-Doherty Lab introduced krypton as an independent check on leak rates. Since these three gases are found in extremely small quantities in Earth's atmosphere, and we knew exactly how much was released into Biosphere 2's atmosphere, their decline over time was a very direct way of determining our leak rate.

The unprecedented low leak rate of Biosphere 2, under ten percent a year, was due to Bill's diligence and the superb efforts to develop new sealing techniques. To appreciate how airtight Biosphere 2 was (given its three-acre size, twenty miles of joints covering 6,600 glass panels, and its vast stainless steel underground liner), a one percent annual leak rate was caused by microscopic holes altogether amounting in size to your pinky fingernail! In contrast, a typical twelve-foot by nine-foot by eight-foot-high office at the Environmental Protection Agency (EPA) minimum standard of fifteen cubic feet of air exchange a minute has an annual air exchange of 100,000 percent which is ten thousand times greater than in Biosphere 2. The Russians with their Bios-3 closed system (the previously most tightly sealed structure in the world), had a leak rate of about fifty percent per year. NASA, at its Breadboard facility at Kennedy Space Center, converted a Project Mercury pressure chamber into a closed system for plant growth for space. They got their leak rate as low as three to five percent *per day* (more than 1,000 percent per year) and stopped there.

A living system, however tightly sealed, must still permit a sufficient flow of energy and information. Earth is a closed system like Biosphere 2, yet it remains open to energy and information from the outside. Earth continuously receives energy from the sun, which permits life to flourish here. If the Earth were denied a source of energy, it would quickly entropize, drop to a level of energy that could not support life, and reach a state of chemical equilibrium like the dead surfaces of the moon and Venus. For Biosphere 2, energy is supplied in the form of sunlight and electricity. The electricity produces hot and cold water for heating or cooling and was supplied by generators (powered by natural gas) and other equipment in an outside Energy Center just a few hundred yards away. Energy was sent out of the facility in the discharge of excess heat. Biospherians sent and received information through computers, faxes, telephones, and video.

The Earth is almost, but not totally, materially closed. In addition to the loss of hydrogen and other light gases from our upper atmosphere, the Earth also receives over a hundred thousand tons of cosmic debris (meteorites, etc.) a year from space. Biosphere 2, through leakage and the export and import of research samples and tools had less than a ten percent exchange rate with the Earth's atmosphere per year. These imports or exports of scientific materials have a negligible mass as compared to the mass of the entire system, so these items are essentially an information exchange. Biosphere 2, like Earth, remained a materially closed system, while open to energy and information flows, and remained so when we had left.

BREAKDOWNS AND REPAIRS

There had been a few problems with the quite numerous pieces of equipment inside Biosphere 2. For example, one type of PVC piping that was installed didn't have a high enough PSI (pounds per square inch). It frequently failed in our wilderness water systems due to the high water pressure, so Laser went around and systematically replaced that type of pipe. Similarly, a few systems were made unnecessarily complex, so we simplified them. This was especially true of some of the computer programs for the technical systems. When Laser and the rest of us started using them it became clear they were not as user-friendly as they should have been. We appreciate the old engineer's motto: "It takes a genius to make a system simple. Any fool can make it complicated!"

A lot of the technical work involved the fabrication of new systems. In the ocean, we started the two-year experiment relying on the algae scrubbers to clean the ocean and marsh water quality. There were sixty of these scrubbers, each composed of a four-by-eight-foot fiberglass box in which two plastic mesh four-by-four-foot screens were placed to facilitate the growth of a community of some forty types of marine algae. Water was pumped from the ocean and marsh and dumped across the screens as if from 'waves' via a plastic bucket which fills and splashes over when full. These algae scrubbers concentrated the functions of ocean algae, which grow rapidly, remove nutrients and carbon dioxide. Every ten days, the biospherians would perform the function of algae grazers by removing the accumulated algae. Then it was dried and stored, along with the weeds as part of the on-going biomass carbon and nutrient management.

We discovered that this system, in addition to requiring up to twelve hours per week of biospherian labor, was inefficient and incapable of doing the job required. It became clear that an additional nutrient removal system was needed, so six months into the experiment, with advice from Dr. Phil Dustan, Gaie agreed to employ a different type of nutrient removal device called a 'protein skimmer'. Skimmers are composed of a fiberglass pipe through which air bubbles. As the bubbles rise against a stream of water, dissolved tannins, acids, and nutrients accumulate on the bubbles and get 'bubbled out.' This is analogous to the foam that accumulates at the edge of a beach from waves lapping along the shoreline. With the advice of aquarist Julian Sprung (Two Little Fishes) and the successful trial run of one skimmer installed by Keaton Fisheries before our entry, we designed and built seven new skimmers for the ocean.

Wisely, Laser stockpiled lots of miscellaneous materials inside Biosphere 2 on the off-chance that we'd have to fabricate some entirely new system. So he and Gaie took stock of the inventory of pipes, pumps, air compressors, air stones, hoses, and other miscellaneous parts stored inside the Biosphere and found what they needed to construct and install the skimmers. After they started to operate, the ocean water immediately began to improve its clarity.

Because the skimmers were so effective, we were able to shut down some of the scrubbers to save labor. The lights from these scrubbers were given to the agriculture team to enhance food productivity. The crew then began an extensive process of making new planter boxes and mounting these artificial lights for supplemental lighting in shaded areas such as in the basement underneath the main fields of the agriculture. Each light was taken out of the scrubber room, washed down to remove the accumulated salt deposits from the continual splashing of salty water, and re-mounted over the make-shift soil boxes. This increased our growing area by approximately 1000 square feet. One of the unique aspects of living in a closed system is that you are always looking for a way to use space and materials more efficiently. It was sort of like a chain reaction—once the ocean nutrient problem was solved and the valuable lights freed up, a whole new project began.

During the second year, Laser and Gaie began the process of re-constructing the marine marsh pumps. Initially, the pumps, which recirculated water throughout the six marsh ecosystems, continued to cause endless problems because the contractor who installed them placed the intakes too close to the sediment. What resulted was a continual intake of bottom sediments, leaves, and some small animals, like crabs, who unsuspectingly were pulled in by the suction of the pumps. Thus, Laser hit on the clever idea of changing the existing pipes into an air lift system. To do this he placed an air stone (just like an aquarium air stone but larger) in one end of the recirculation lines. This created a water current that ran out the other end of the pipe thus creating a flow of water without any mechanical action. All the six marsh ecosystem currents were thus driven by air! If we had thought of this years before, we would have lost far fewer fish and crabs. The loss of such animals was an inherent problem in meshing pump technology (for necessary water flows) with ecological systems. This method would have also saved us many hours of repair work.

General maintenance was carried out on the computers, sensors, analytical equipment, and video systems. Roy and Linda helped Laser keep the many critical sensors calibrated. Frequently, Roy could be heard on the radio with Sherri Burke, the sensor specialist at Mission Control, as they calibrated the agriculture air temperature or ocean water temperature. Linda spent an hour a day working with Gary Hudman (also in Mission Control) on the air handler valve controls which proved to be a recurrent problem. Taber kept the analytical and sniffer systems operational. All computers, analytical machines, and sensors have their own stock of spare parts and tools for repair, maintenance, or calibration.

Laser managed the video systems inside the Biosphere which involved a communications-conferencing system, remote cameras for viewing the wilderness and agricultural areas, entertainment (there was a set up for piping music and videos into Biosphere 2), and media interviews. The media events caused problems because they were complex—especially those involving satellite uplinks. Each required a camera that was set up to send an image out to satellite trucks that relayed the images to interviewers who might be in New York, London, or Tokyo. Interviewers wear remote microphones to feed their voices to the satellite truck, while providing a two-way communication back to the Biosphere through an ear piece attached to their hand held radios. Setting up for a media event often took Laser two full days: one for setting up and one for managing the link-up and dismantling the system after its successful operation.

Other technical tasks completed during Mission One were more mundane but equally essential. Laser was on call for everything that needed fixing. Instead of calling the plumber, carpenter, TV repair man, electrician, or dishwasher mechanic we called on him. Instead of going out shopping for the latest gadget or light bulb, we asked Laser to either fabricate it or escort us to his spare parts stores where we found what we needed. During the course of the year all of the biospherians became more involved with the hands-on maintenance and repair of their areas. At first, some biospherians couldn't even figure out if an electrical device had failed due to a tripped breaker or was actually broken, But, slowly the calls for help with every little glitch started to decline, and several of us who weren't mechanically inclined actually began to fix our own problems. However, all along we knew that if we failed, 'Mr. Fix-it' could be called on as a final resort.

Aside from not having to pay huge repair bills or wait for the arrival of a specialist to come and repair equipment, there was another aspect of maintenance unique to our system. We couldn't afford to assume something was useless and toss it into the garbage can if it was broken because we didn't have the option to go out and buy another! We had to make it work if at all possible. Electric kettles were continually fiddled with to make them work, and one of the electric blenders was epoxied back together after it was dropped on the floor and broken into several pieces. Although battered and worn, it kept on blending sauces and soups.

ALARM WATCH

Just like a ship, Biosphere 2 required that someone always be on watch to make sure that all systems were operating smoothly. Also, there was always one person from Mission Control monitoring the biospherian channel on a two-way radio for automatic alarms or requests for assistance from the biospherian watch of the day. The inside watch included a night-time check where the biospherian toured the entire system from rainforest to agriculture. Air temperatures were noted and specific checks of problem areas like the marsh and ocean water flows and waves were made. In addition, the watch double checks to make sure the kitchen stove is off. This check was introduced after the first few months because more than one of the cooks was negligent and left the daily slops for the pigs cooking on our electric stove. Our pigs were dainty eaters who only touched their slops if they were nicely cooked. Luckily, the danger was averted by crew members who woke up in the middle of the night smelling smoke!

Biospherians always carried two-way radios for communication with each other no matter where we were in the Biosphere. The radios were also useful for contacting outside staff when needed. In addition, it was a safety measure in case of medical, fire, or other emergency situations. At night the radios were placed in recharger units close to where we'd sleep so we could hear an alarm if one should sound. For the watch person on call, the radio is the primary method used to alert others to possible problems. The radios were connected to a computer alarm system that had an 'alarm 1', 'alarm 2', 'alarm 3', 'leakage', or 'pressure' radio messages. These alarms would be broadcast over the radio in a robotic voice until the problem had been solved.

Laser was the only one who did not carry out a twenty-four-hour watch. The other seven of us took watches in rotation—we always had our watch on the same day of the week. But as the Emergency Captain, Laser was responsible for responding to all legitimate alarms that couldn't be solved by the watch person. Like most systems, ours would occasionally sound a false alarm. One time when Gaie was on watch the radio went absolutely berserk. All of a sudden there were multiple alarms coming over the radio in rapid succession. This had all the other biospherians in peals of laughter and several called over the radio, "Gaie, it's your lucky day, you hit the jackpot!" Luckily, it turned out that someone in Mission Control, trying to re-set the program, triggered a series of false alarms.

A couple of weeks after closure, Mark Nelson called Laser at about three in the morning to report a recurring high-pitched noise. Thinking it might be a fire alarm, a mechanical alarm, or an indication that something was going to blow its lid, he quickly telephoned Laser for his opinion. After a few seconds as Laser listened to the noise over the phone, he concluded with a great laugh of relief that it was a cricket which had somehow gotten into a pipe or wall of the habitat!

On another occasion, Laser's voice came over the radio, "Emergency, emergency, all hands to the desert near the scrubber room. Bring a stretcher and medical kit, Mark has fallen down the stairs and broken his leg." All of us went into emergency mode and one by one responded over the radio that we were on our way. Gaie grabbed the stretcher, Roy and Taber the medical box, and Sally manned the radio and telephone to keep the outside posted. The Mission Control watch called security, the on-site paramedic, and an ambulance to stand by just outside the savannah airlock. When we arrived at the scene of the accident a couple of minutes later, we found out it had been a drill! Leave it to Laser to keep us on our toes.

In the event of a real accident, four of the biospherians had been specially trained for emergency response, and all eight went through a series of drills before closure to help us prepare for potential accidents like electric shocks or fires. Roy, as medical officer, was always ready to handle any accident. The outside watch was responsible for coordinating the emergency staff on the outside so that if an intervention were needed, they'd be ready. A paramedic was on site during the day and available at night, if required. The site also has fire-trucks to supplement the Biosphere 2 system.

In addition to medical drills, Laser also ran unannounced fire alarm drills. Fire would be a major problem in Biosphere 2. Even a small fire might necessitate the crew leaving and the atmosphere being flushed out since the smoke would quickly pollute our small system. To handle this potential problem, we installed an extensive system for fire prevention and response. It included smoke detectors throughout the Biosphere along with many fire alarm boxes. An automated system sounded a loud alarm throughout the Biosphere in case a fire or smoke alarm went off. The central fire alarm control panel in the command room displays where the alarm originated. Once the alarm sounded, the biospherian nearest to the command room heads there to inform everyone else of where the problem originated. Those trained in fighting fires had quick access to fire-fighting equipment, which was located at half a dozen locations.

Because the glass in the space frame was so strong, we had to install some conventional glass panes that we could easily break in case of a major fire. The Biosphere 2 panes of glass were stronger than car windshield glass. Even a strong impact was unlikely to cause a break through the pane. During construction, a glass pane fell from the lifting crane some forty feet to the ground and it didn't even show a crack! If one of these panes should break, there were plywood covers we could place under the glass to minimize leakage while an outside crew installed a new pane of glass.

A near emergency occurred on the night before we entered. Thousands of people wandered around the grounds eating and drinking under huge, festive white tents while laser-light beams danced on the glass and space frame Biosphere under a blue, starlit night. Fireworks erupted in rainbows of color. Gaie was inside the Biosphere walking down the spiral habitat staircase when she smelled smoke! She ran towards the wilderness area where it came from. Calling for help on the radio, Bill Dempster responded that he was the one who had generated the smoke by trying to incinerate algae in the scrubber room. This was one of the ways we considered for recycling the nitrogen, carbon, etc. stored in the algae tissue grown on the algae screens. Bill was using the last night to experiment with different ways of using this incinerator. He spent the rest of the night flushing Biosphere 2's air (which had been scheduled anyway), and the smoke was quickly gone. We were going to start the two-year experiment with fresh outside air to track exactly what changed from day one, September 26, 1991, as Biosphere 2 began its divergence from Earth.

WITH A LITTLE HELP FROM OUR FRIENDS

One of the unique aspects of our technical systems was that we could get advice and help from people on the outside. For example, support staff in Mission Control checked the computer readouts of system status and spotted problems we may have overlooked on the inside. We could also call up the companies that made our equipment to check on the best ways to undertake repairs or to enlist their assistance if the equipment failed to perform as specified. Engineers were eager to help when they understand the call was coming from inside Biosphere 2. Companies were also highly motivated to get their equipment working up to specs when they realize that this little world was dependent upon it.

With the video system we have inside Biosphere 2, we can link up with specialists to look at problem equipment or parts. They could also talk Laser through a repair with which he was unfamiliar. A good example of how creative this repair and engineering at a distance could get is the assistance Taber received when the liquid nitrogen plant began to malfunction. This equipment extracted nitrogen from our air and liquefied it for use in the analytical laboratory. It was made by a firm outside London who helped tailor it for Biosphere 2's needs. When we began experiencing problems with it after closure, the engineers set up a similarly modified machine in a room so they could duplicate the conditions it was subjected to—high carbon dioxide and humidity. They checked one possible cause after another. Finally, they discovered that the SF_6 gas we'd spiked the Biosphere 2 air with was causing the problems. They quickly worked out a solution using their machine outside London as a guinea pig. Luckily, it also worked for the device inside Biosphere 2. This was a perfect example of the power of global communications. The trickiest part of the repair by long distance was the eight-hour time difference between Arizona and England.

Biosphere 2's challenges attracted some of the world's best and most innovative research organizations, corporations, engineers, ecologists, and geochemists. How often are you offered the chance to build a rainforest from scratch—from the underlying stainless steel and concrete on up? Or to recreate a miniature living ocean, moving hundreds of species of coral reef from the Caribbean and Yucatan, which can scarcely tolerate even a few hours without the right water quality and movement, light and temperature? Or to engineer everything from ocean waves and tides and savannah stream flows, to making a virtually airtight structure with over three acres of glass roof, and twenty miles of joints? And all this must be accomplished in a greenhouse which recycles its air and water under a blazing desert sun in summer, snow and wind in winter!

Sally holding a newborn goat. The domestic animals were a source of entertainment, affection, and contributed a small portion of our diet.

CHAPTER 9

Animal Tales

"The whole project seems to me like the star that's up in the air, and we look at it to see if man can survive on another planet. So the whole project is like on a table and you look at it from different angles and you look at it from different dimensions."

– **Dan Old Elk,** Crow tribal medicine man at the
Biosphere 2 entry ceremony, September 26, 1991

DOMESTIC ANIMALS

WE SPENT TWO YEARS living with animals—and we don't mean cats and dogs. When we first entered the Biosphere, we had no idea just how important a role animals would play in our lives during the mission. Certainly the animals had been carefully selected after years of experiments at the SBV research and development center. The domestic animals were chosen for their productivity and ability to survive on the diet we have available. They ate the parts of plants that humans cannot digest such as sweet potato and peanut greens, or sorghum stalks. They turned this foliage into milk and eggs for us to eat, and their wastes were valuable for making compost, our renewable source of soil fertilizer. The wilderness animals were selected for the roles they play in the food chain. In addition to their functional roles, the animals played a vital role as companions, and even entertainers, for the biospherians.

This was particularly true of our domestic animals. We went into the experiment with three types: African pygmy goats, Ossabaw feral swine, and a chicken that was a cross between a Jungle Fowl and a Japanese Silky breed. We chose pygmy varieties because their small body size meant they were more efficient food-utilizers, so that for the amount of fodder available we were able to support greater numbers of individuals than would be possible using larger breeds. Of course, these animals provided only a small amount of food in our diets but more importantly they were an important component of a farm system, They helped process vegetable waste, as well as provide humor and companionship to us all.

Goats give nourishing and tasty milk, but everything is miniature compared to a cow. From their tiny teats, we only got about forty ounces of milk per day from our four goats. Not a lot, but milk was a wonderful staple, providing the raw material for sauces, cheese, ice cream, milkshakes, and other greatly appreciated delicacies.

151

Our chickens were like the original wild chicken—a hardy breed that forages for insects and eats just about anything they can get. Their highly bred domestic cousins can only subsist on a diet of commercial chicken feed and have lost all their innate ability to brood and raise chicks. We did not want to use incubators in the Biosphere, so we crossed the jungle fowl with the Silkies because the Silkies are such good mothers, regularly brooding and raising chicks.

Obviously, animals need a comfortable environment in order to remain healthy. The African pygmy goats have a mountain ancestry so we provided a variety of wooden boxes to climb; the pigs were given boxes of soil to root in, water tanks in which to wallow, and any number of surfaces for scratching themselves (which seems a source of inexpressible pleasure to a pig). The chickens had overhead roosting perches, straw-lined laying and hatching boxes, and plenty of soil and mulch to scratch their way through. Roosters even had rivals in adjoining pens they could attempt to dominate. Watching them eye each other, crow, and leap up against the wire netting separating them was quite a spectacle. As a matter of both aesthetic pleasure and functional efficiency, the pens had an abundance of plants growing in window boxes and containers. They were mostly fodder plants or yams that formed a green arbor over the top of the pens, or bananas whose leaves were periodically cut for feed. The plants and wide spaces of the animal pens gave the animal bay the feel of a tropical garden and farm.

Anyone who has lived near animals knows they can vary widely in personality. Our four milk does were very finicky. Having been treated like prima donnas from day one, they would accept nothing but the very best. They made it clear if the snack used to lure them to the milking stall was unacceptable. Sally and Jane shared the milking duties. Each morning they made their way to the milking pen with the milk buckets, a saucepan with which she initially catches the milk, and a basket of goodies—usually some sweet potato greens or bean hulls—to keep the goats busily munching as she milked.

The goats have a milking order, and there was much pushing and butting at the gate of the milking stall if someone tried to go out of turn. Sheena, the largest of the does, insisted on being first. Generally, she stood quietly while the milk squirts into the bucket, but if the sweet potato greens were a bit on the limp side, or she did not think there was enough food in the trough, she gave a swift kick in the direction of the milk pan. Sally and Jane grew wise to this over the months and were always ready to grab the pan quickly if they saw a kick coming.

Milky Way, the youngest doe, was terribly nervous at first and it took her about a year to settle in. She had previously been raised almost strictly on alfalfa hay and it took a while before she recognized Biosphere 2 fodder as acceptable food. Once terrified of humans, she grew confident, bounding up to the milking stand and no longer minded being stroked. Stardust, on the other hand, was the Mack Truck of the goat community. Nothing fazed her as she went about her business of eating anything she could.

Vision butted and bullied the other goats and often stubbornly refused to come and be milked. She always seemed put out by the whole affair of milking and unless the snack was particularly delicious would often stand in the stall and disdainfully refuse to touch it, her nose in the air for the whole procedure. Generally, the only thing that pleased her was fresh banana peels, the most sought after delicacy in the goat world.

Buffalo Bill, the buck, loved to have his head scratched and went up to the humans entering his pen and rubbed up against them begging for attention. If the female goats were in heat and he was unable to get at them, he ran around butting everything in sight, making growling and spitting noises. He was very intelligent and frequently succeeded in releasing the latches on the pens. We started to call him 'Houdini' since it seems he was able to get out of anything. Eventually we had to wire up the doors on his pen. All the goats, once their gates had been opened, went straight for the feed bins. If no human found them, they ate until they were practically immobile—rather like a human after a traditional Thanksgiving dinner. When they were finally discovered, they were so full that they were barely able to waddle back to their pens.

We soon became intimately familiar with goat and pig health and family life. Usually the goats bear twins, but they occasionally have triplets. These births were not always without drama. Our first goat births occurred on the same day in the middle of November 1991. First, Stardust gave birth to twins before anyone arrived for the morning feeding. Then later in the morning, Vision went into labor, but the kid wasn't positioned correctly. Even with help from the biospherians it couldn't be correctly aligned. The drama went on for some hours until Dr. Barbara Paige, the vet from Tucson who had worked with us on our animal program, came out in the early afternoon with a 'demonstration kid' (one of the new-born kids from the BRDC). She tried to show Jane how to grab the kid while it was still in the uterus. She explained that sometimes the kid will just be too big, and it may not be possible to save it. But the critical thing is to save the mother's life. Unable to help, Jane called Laser in to help—they had both worked on a farm in Australia before, and Laser's hand was stronger. He was finally able to pull the kid out, but it was stillborn. Luckily, they were able to save the life of Vision.

In a curious twist of animal behavior, the twin kids started suckling from both Stardust, their biological mother, and Vision, their surrogate. Perhaps since both does went into labor and had milk at the same time, they weren't able to tell whose kids were whose. Since Vision seemed to be the better mother, we assigned the kids to her—to the evident satisfaction of all parties.

Sheena injured her leg—we don't know how she got the original wound, but by the time Jane noticed that she was limping, the infection was very deep. Jane and Taber regularly cleaned the wound and dressed the leg in an attempt to nurse her back to health. At first they administered tranquilizers to make Sheena stay still enough to treat the sore. Sheena grew to love these shots, and leaps up on the bench that Jane used as a medical table to eagerly await the injection. Our goat turned into a junkie! Later as the wound healed, the shots were no longer necessary. Sheena, with a slight limp, resumed her self-appointed role of 'queen of the goat pen'.

The two adult pigs, Zazu and Quincy, were captured on Ossabaw Island off the Georgia coast, an area where Gaie spent a lot of time as a girl. The pigs there can be quite aggressive. Gaie remembers her parents telling her not to excite the wild pigs in any fashion because they may charge, and if they did she should hide behind a tree. She held them in great respect. Zazu and Quincy were anything but wild, and along with their seven piglets they provided never-ending entertainment. The piglets loved to nap inside the black plastic watering bucket, all piled up with legs and snouts sticking out everywhere.

Zazu and Quincy slept side by side in one of the wooden boxes. We called them Mr. and Mrs. Piggy. Quincy, who was the smaller of the two, was hen-pecked by Zazu. She gives him a sharp nip at feeding time if he got too pushy going after the food. One day Linda discovered that if she scratched Quincy behind his ears, he promptly goes into a semi-comatose state and drops at her feet in complete pig ecstasy. A few days after closure, Zazu led the entire crew on a wild hog chase when she managed to escape into the IAB. Having found pig-paradise in our vegetable plots, she was not going to give up easily. Frantic biospherians ran through the fields trying to cause as little damage as possible while they headed off the rampaging pig. Eventually she was returned to her pen, but not without a struggle.

Delightful though they were, the pigs began to pose some problems. We originally worked with Vietnamese pot-bellied pigs, very small creatures that did not eat a great deal. They have such a gentle temperament that they became popular across America as pets. It was this popularity that eventually prevented us from taking them into the Biosphere. Animal rights activists, unable to view the animals as part of a farming system, begged us not to take them into Biosphere 2 and eat them. We decided to drop the pot-bellied pigs, but this meant we had to find a last minute replacement.

The Ossabaw feral swine are not as small as the pot-bellied pigs. They eat a fairly large amount of green leafy material, which was all we had available to feed them, but they were not really thriving. The pigs much prefer a starchy diet, but we couldn't afford to give them any of the starches since these were a crucial part of the human diet. Also, just collecting fodder every day to keep them going was taking hours of crew time. Eventually, we faced the sad fact that we could not support the pigs in our system. They were butchered and put into the freezer to provide special delicacies for Sunday dinners and feasts for many months to come. As our agriculture area matured and supplied more fodder, we might have been able to consider introducing another breed of small pig into the system, but for the rest of the two-year experiment we had to make do with chickens and goats.

For the first year, our chickens produced few eggs. This isn't too surprising since they were at the end of the line as far as food distribution was concerned. With chickens, you get out what you put in. They got some kitchen slops, though not much, since we usually ate everything on the table. They also got the cockroaches and pill bugs that we trapped, the worms Laser raised, some azolla from the rice paddies, and grain stalks from grain processing. They looked healthy, if a little skinny, but plentiful egg laying was just more than could be expected of them on that diet. In the second year, when we raised more food in the IAB, Sally could give them scratch (the scraps left over when grain is threshed) for short periods in large doses and more trapped insects. The chickens responded by laying a few more eggs. These eggs were highly prized items for holiday and birthday cakes and crêpes.

By the second fall, our first three chicks hatched. When they were old enough, they were moved to the goat pen where they ran happily with the goats and pecked at their slops. The chicks were also occasionally seen hitching a ride on a goat's back. Totally tame, they could be picked up and stroked by humans without ruffling a feather. In the mornings, when Sally was milking, they came for a visit to see if any of the goat goodies were tasty. Then they stood in the feeding trough and pecked away alongside the goats. Otherwise, they'd be happy to just stand around and cackle sociably.

In order to keep the goats in milk, we had to schedule their breeding so that kids are born at periodic intervals. Once they reached a certain size, the goat kids had to be butchered. Biosphere 2 was able to sustain only a limited number of animals. Laser was trained in the use of a stun gun, a method chosen because it was extremely quick and humane. The animal carcass was then hung and butchered, and the joints of meat were carefully packed away in the freezer in meal-sized packages. Animal organs like liver and kidney were cooked up fresh for the evening meal, a treat greatly enjoyed by all the crew. Every last bit of animal was used. The pig's head and hooves were used to make posole and soups, and Sally spent several evenings in the processing room making her first blood sausages.

GALAGOS

Wild animals were also an integral part of our world. The stars of the animal world in our wilderness are the galagos, small primates commonly called bushbabies. One of the deciding factors to include the galagos inside Biosphere 2 was that they'd provide some companionship for the humans. Even though they were raised in captivity at the Duke Primate Center and later spent time in our research greenhouses, we expected the galagos to adapt fairly quickly to being 'wild' again. In the years before closure, we enjoyed watching the galagos as they were re-introduced to the world of trees, fruit, and insects. Previously, food had only come to them via scientists in white lab coats. So what would unfold when they had access to the wilderness biomes of Biosphere 2?

Four galagos were introduced into the Biosphere. Oxide, an adult male; Topaz, an adult female; Opal, the subordinate female; and William Kim, a baby female galago born in the research greenhouses to Topaz. (The baby was named after the writer William S. Burroughs who urged us to include a small prosimian in the Biosphere.) There was already bad blood between Topaz and Opal before closure. Topaz badly injured Opal shortly after her birth in the year before closure. The female brawling may have stemmed from jealousy regarding Oxide, or perhaps it was a simple assertion of hierarchy. Topaz was the alpha female, Opal the lowly beta, and Topaz didn't let Opal forget it. We hoped that the larger territory the galagos commanded in Biosphere 2 might lead to some kind of peaceful co-existence.

The question of whether they'd confine themselves to the rainforest, which is rich in food and similar to the habitat of their native Africa, was answered a few weeks after closure. We were enjoying our Sunday dinner when the galagos were spotted dashing across the overhead pipes that ran through the upper savannah. They stopped as if to listen in on our conversation, then continued on their way. In time this came to be known as the galago highway because we often saw several galagos heading south in the direction of the desert or north back to the rainforest.

The first inkling we got that the Topaz/Opal drama was continuing was when a night-watch spotted Opal in the technical basement. We were surprised because we anticipated that the galagos would stay in the wilderness biomes overhead. Galagos are both nocturnal and arboreal; it is thought that they instinctively stuck to the trees and disliked being on the ground. In their native Africa, coming down from the trees would expose them to predators.

While performing a nightly check of Biosphere 2 in late October, Mark was startled when he entered the desert and heard the loud hooting of a galago. A survey of trees and space frame revealed Topaz in one of the tallest of the upper thorn scrub trees. Mark was answering her vocalizations, hoot for hoot, when he became aware of further movement south. To his great surprise, Mark saw Opal scrambling over the desert rocks. It became clear that Topaz was intent not only on Mark, but on her female adversary. Going to the ground was Opal's way of showing subservience to the alpha female.

Biosphere 2 is a galago's paradise. Where the trees end, there were the monkey-bars: the space frames. Galagos are small (an adult weighs just a couple of pounds), and their food needs are modest. But with a long tail and superb jumping ability, they're capable of prodigious leaps and can cover distances quickly. In Biosphere 2, where the trees were near the space frames, they leapt from one to another without missing a beat. On one overcast afternoon a few months after closure, while pruning back tall grass in the upper savannah in our carbon dioxide battle, Mark and Linda were riveted by the spectacle of Kim learning the ropes in her new world. She must have been asleep in a tree above the bamboo belt that separated the rainforest from the ocean, and evidently, she was awakened by the human activity. They saw her scamper along a space frame that ran just under the glass from east to west. Just a couple of feet from a connecting space frame node, she slipped and managed to grab onto the space frame by her powerful hind legs. While they watched entranced, Kim made a couple of efforts to swing around and reach the node with her forelegs. No go. Was she in trouble? There were no vocalizations, no frenetic activity. But Linda never heard of galagos hanging suspended upside down either.

They speculated that the galagos, normally active at night, were used to far drier and easier to grip space frames. During the day, and especially near the overhead glass, the space frames are wet and slippery, covered with condensation. While Linda and Mark lay on their backs intently watching Kim through field glasses, they began to toss rescue scenarios back and forth—tall ladders, biospherians with safety lines climbing the space frames to reach her some thirty-five feet above the ground. There was a lot of vegetation below the space frame so even if she slipped she might be able to grab an acacia branch (ouch! the thorns!) to break her fall. Finally, some twenty minutes later, a few quick swings around the space frame brought her nearer to the node, and with an assist from her tail, she was back upright. A split second later she decamped onto a nearby Leuceana tree—as if she only trusted the natural branch now that she'd seen how slippery this steel tree proved to be. The sun was setting. Mark and Linda let out a collective sigh of relief—no need for the rescue team today.

In mid-October after Roy spotted Opal in the tunnel that leads to the south lung on his night watch, Linda went down to lure her with a banana. Linda nursed Opal after she'd had her row with Topaz before closure and was closer to her than anyone. Sure enough, step after cautious step, Opal finally came within reach of the banana. Linda was concerned that perhaps Opal was lost in the technical basement of Biosphere 2 and couldn't find her way back to the biomes overhead. Linda placed her back into the galago cage in the lowland rainforest, where she could be fed for a while. Kate Izzard, the galago expert at the Duke Primate Center, counseled that Biosphere 2 should be large enough for two females to stake out different areas. So early in November, the cage door was left open for Opal to make a second attempt at life in the wild. While Linda watched, Topaz came in for her food set out on the cage top, and stood guard from a nearby tree glowering at Opal who crouched fearfully inside. When dawn broke the next day, Linda took Opal out of the cage. Opal had an hour or so to find herself a comfortable sleeping spot before the morning light signaled her it was time to call it a night.

After that, we became accustomed to seeing Opal in the basement. For a while, she slept near the ceiling of the room that leads up to the room with the algae-scrubbers. Linda put out her chow in the technical basement to ensure that she had access to food to supplement what she was able to forage upstairs.

One afternoon, while working on calibrating sensors, Roy discovered Opal curled up asleep inside one of the boxes that relay signals from our video cameras!

When Linda reached Roy, Opal had partly awakened, and was regarding them through sleepy, half-closed eyes. Since the box contained no live current, she was left to enjoy the rest of her sleep. Later, Gaie saw Opal by the ocean, sleeping in a plastic bin on the diving platform!

Early in March 1992, we celebrated the birth of our first Biosphere 2 galago baby. Oxide was very protective of her, and it appeared she was, by the time we saw her, at least a few weeks old. But since the gestation period for a galago is about four months, she was very likely conceived inside Biosphere 2. Although this did little to appease the interest of those journalists who persisted in asking us whether a human baby was likely during the two years, it marked a milestone in terms of the acceptability of our artificial world by our evolutionary relatives.

But a tragic accident occurred later that March just a few days before we celebrated our first six months inside Biosphere 2. William Kim had evidently also taken to exploring the technical basement, and died of an electric shock when she touched a power transformer box with a small opening virtually invisible from the floor below. Deeply saddened by the loss, Laser and Linda spent time the next few days meticulously checking and safeguarding against any other dangers that the tiny fingers of our galagos might somehow encounter since it was now clear they possessed the curiosity to roam the parts of Biosphere 2 most remote from their rainforest habitat.

In the meantime, Opal's drama was far from over. Despite Opal's retreat to the dungeon, the technical basement, Linda found her in the summer with an ugly-looking scalp injury, probably the result of a scrap with Topaz. Fortunately, Roy's X-rays showed no bone damage. We kept her in a large cage in the medical room for some weeks while we administered antibiotics and cleaned her wound regularly. Then we dismantled the galago cage from the rainforest and moved it into a shady corner of the orchard (on the agriculture side of the Biosphere) so that Opal didn't have to endure harassment from Topaz. But when we made her new room ready, Opal found a tiny gap in the cage's overhead mesh and announced her escape by hooting from the giant banana trees in the orchard. By the time the hole was located and patched, Oxide joined Opal by slipping through loose screening that divides the savannah from the orchard.

They refused to be lured back in, so then came the fun of figuring out how to trap them. We finally came up with a long string attached to the door of the galago cage. Baited with ripe bananas and monkey chow, biospherians took hourly turns discreetly seated in a wooden chair some twenty to thirty feet away. The trick was a quick release of the door while the galago was inside snacking. It worked. Oxide was trapped and returned to the rainforest. Opal was reintroduced to her tree-shaded cage.

But no sooner did we think order had been reinstated than the familiar sounds of night hooting from orchard trees and the IAB balcony told us that the galagos from the wilderness were determined to have the run of the human areas as well. We'd been afraid of damage that galagos might cause if they have access to the IAB, but careful checks revealed they limited themselves to occasional nibbles at the ends of nearly ripe bananas. They prefer heights, and showed no interest in clomping through the agricultural plots and rice paddies. It seemed we could live with the arrangement, and in fact we have been delighted by the occasional galago conversation late at night when went out for a stroll and leaned over the IAB balcony. The only place that our irrepressible galagos hadn't successfully made part of their domain was the human habitat. We often joked, though, that one of them may turn up for a seat at the dinner table.

Thereafter, Opal seemed content—safe from her female nemesis. She was slated to rejoin our growing galago colony in the research greenhouses after the two years. Like us, she would have her own unforgettable memories of being a participant in the first two-year closure of Biosphere 2.

STOWAWAYS

There were several types of animals and insects that were not deliberately introduced into the Biosphere, stowaways who refused to leave. In the ocean biome, there were octopi that were transported from the Bahamas along with the algae reef rock who remained hidden for at least a year before being noticed. The first one was caught before closure by Laser, but it was not an easy task. He had to wait silently in the ocean at night for it to come out. Then, during closure, Gaie noticed that the Indo-Pacific giant clams (accessions from the Philippines) were being picked clean. Suspecting octopi, Laser dove again and caught three more! That was the last of the octopi, a magnificent animal that we were sorry to remove, but our ocean could not support its voracious diet and loss of all the giant clams.

In the terrestrial biomes, there was a large curved-bill rock thrasher who hung out in the savannah, thorn scrub, and desert. At first we tried to snare it and remove it, but it was wise to the bird nets we'd used to catch most of the English sparrows

Laser designed this well-equipped workshop to repair or fabricate new replacement parts and keep the extensive technosphere of Biosphere 2 functioning without relying on importing parts and supplies from outside.

Dr. Roy in the medical lab. The Biosphere 2 experiment and its novel environmental conditions, such as reduced oxygen and our low-calorie/nutrient dense diet, produced cutting-edge scientific research into human metabolism and physiology.

Mark, Linda, and Jane smile during a harvest of peanuts. They were an important source of fat in our diet and their greens were relished by our domestic animals.

Jane is feeding peanut greens to our African pygmy goats. The domestic animals in our small barnyard area were a source of joy, companionship, entertainment, help in recycling inedible crop wastes, and supplying special treats for the diet—milk, eggs, and meat.

The south-facing agricultural basement included constructed wetlands for wastewater treatment, planting boxes for rice, and rice seedlings. Mark and the crew added planters wherever sunfall was not being used to capture atmospheric CO_2 and grow more food.

Abigail Alling

Feasts were collaborative—and usually all-day—events; cook in the morning as to enjoy and party later! Sally, Mark, and Jane work on their special dishes in the ultra-modern electric kitchen. Gas appliances would have been a source of air pollution.

Linda speaking with visitors about the research conducted within the facility. During Mission One, Biosphere 2 attracted over 250,000 visitors a year, making it one of the largest tourist destinations in Arizona—second only to the Grand Canyon.

Gaie uses a video link with Mission Control. Video media hookups and conferencing facilities for meetings between inside and outside scientists and engineers happened in the Command Room, where each biospherian also had a computer and desk. Several large computers also displayed real-time sensor and other data that was being collected.

Birthdays were celebrated in special locations, and this one is on the beach! Vacuum pumps produce gentle lapping waves that keep the ocean water circulating, critical for coral reef health.

The crew gathers for breakfast and morning meetings to plan the day's work and research tasks. We reviewed critical data from our sensors, like CO_2 levels, studied weather forecasts, and made weather requests: "Can we have a 30-minute rain at 4PM in the rainforest?"

Taber and Jane rehearse on the IAB balcony to perform a new song during an arts festival where music, art, poetry, and videos were shared between Biosphere 2 and participating outside artists.

The crew celebrate completing the Herculean task of cutting back the grass in the savannah biome.

Dr. Oleg Gazenko (left), Director of the Institute of Biomedical Problems, Moscow and John Allen, inventor and Director of Biosphere 2, exchange biospherian handshakes with Mark.

Sally was the master baker of celebratory biospherian cakes. At the anniversary of the end of our first year inside, she and Laser greet visitors at a viewing window.

A night view of Biosphere 2. The buildings on the left make up the Energy Center, the two geodesic domes cover lungs, and the habitat tower alongside rainforest spaceframe are the tallest structures.

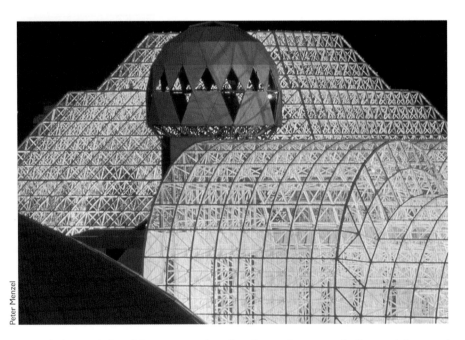

The human habitat library tower, a barrel-vault space frame over the farm and the stepped pyramid of the Biosphere 2 rainforest lit up at night.

before closure for release outside. In addition, the thrasher had a lovely, plaintive song. Because it was not doing much harm, and since the birds that we deliberately introduced had not done well, we eventually decided to leave it. It was nice to be able to hear bird songs in our wilderness, even if it was not exactly the bird song that we had originally planned on.

The same does not apply to the IAB where some sparrows were competing directly with our stomachs. There, a large flock of sparrows moved in during construction and lived happily on what they could steal from our grain crops. We tried everything we could to trap them and get them out. We caught the majority in nets before closure but three very shrewd birds evaded us. One by one they died off until one lonely sparrow was left by the second year. We gave up trying to catch it. We often wondered how it felt, being the only sparrow left in its world.

Mice lived in our basement, and probably all over the wilderness biomes though they were not so visible there. In the IAB basement, we trapped large numbers of them because they could really become pests eating our precious grain supplies. We were careful never to leave any grain where they could reach it. We also trapped large numbers in the habitat area. They seemed to have free run of the habitat via the electric conduits. We tried various different baits. The trick was to use something that they could not remove from the trap without triggering it. Sticky banana mixtures seem to work well, but then the problem was keeping the cockroaches away from the bait that was intended for the mice. Yes, even in Biosphere 2, humans were engaged in their age-old battle against the mouse and the cockroach.

It seems from our experience that interaction with other members of the animal kingdom is an important part of life for human beings. Toward the end of the second year we began to evaluate whether or not we should keep galagos in the system since it would still be some time before the wilderness was mature enough to support them. Everyone agreed that it was worth the trouble of taking care of them simply for the pleasure of watching their antics. The same applies to the domestic animals. We could have devised a diet completely free of animal products and found some other way of composting our inedible plant material, but the domestic animals enriched our lives as well as our diet.

*Measuring the growth of trees and grasses is an important part of
the ecological research conducted inside the facility.*

CHAPTER 10

The Three-Acre Test Tube

"Biosphere 2 because of analogous properties to Earth, provides the opportunity to investigate the potential implications for the global carbon and oxygen budgets that may result from climate forcing and a changing atmosphere... The holistic perspective provided by Biosphere 2, so necessary to understand system level responses within the enclosure, will help shape the emerging interdisciplinary approach to understanding Earth."

– **Dr. Victor C. Engel**, US Geological Survey, and
Professor Howard T. Odum, University of Florida

A UNIQUE LABORATORY

Geophysicist Keith Runcorn was a Fellow of the Royal Society of London who has made significant contributions to the science of plate tectonics, helping to confirm that continental drift is a reality. At the Environmental Symposium held at SBV the day before closure, Runcorn pointed out that while physical scientists have become used to the power of advanced technological tools, scientists who study the environment have very few such tools. Moreover, ecological studies often suffer in the competition for scientific funding, despite the obvious importance of the global environment. This made Biosphere 2, because of its scale, complexity, and data production, a unique laboratory, the first of its magnitude and kind in the arena of life studies. We could not know in advance what we'd learn by working inside it, but as Runcorn pointed out, new scientific tools let you see what has never been seen before.

Our two-year mission was, among other things, a shakedown voyage where we'd learn to operate an entirely new kind of scientific laboratory. Since we were simultaneously carrying on research activities and learning to use the facility, the border between research and operations was crossed back and forth many times in the first two years. For example, data about water quality and air composition was collected for health and safety reasons, but it also provided new information on how these cycles work. In the agriculture system, there was almost no division at all between operations of the facility and the research about sustainable agriculture and chemical-free integrated pest management. To a great degree, the operations were the research, and the research informed the operations. Our primary intention

163

was to observe the unexpected and take advantage of our opportunity to study what is radically new.

There were times when we argued at the breakfast meetings about how much time within the parameters set by mission rules we should allocate to research and how much to needed operations. At one point in the first month, Laser erupted, "Is there going to be a split here: scientists versus technicians?" There were several biospherians who at first were quite unhappy that we didn't have more time for direct research. We all agreed to participate in the first mission knowing full well that all of our time might be used dealing with operational necessities. After all, there are bugs in any new system. But the division of time for research vs. operations was a key issue and one that was repeatedly reviewed and discussed. Without Biosphere 2, there could be no study of biospherics, hence our number one objective was to keep all systems operating. But without research, the long-term purposes of the Biosphere would be diminished.

As we became increasingly efficient at operating Biosphere 2, more time became available for pursuing specific research. By the end of the first year, there were over sixty scientific research projects underway. Although there were only eight of us inside, there was the SBV research staff outside to whom we reported, plus we had hundreds of scientific and technical collaborators, ranging from consultants, friends, to those who simply had interesting suggestions to make. Scientists on the outside exchanged information through computer, telephone, and video. We used state-of-the-art computers, sensors, and multimedia facilities to enable SBV staff and outside scientists to assist us on an ongoing basis in areas of special research interest such as gas cycles, soils, and physiology.

To provide information for both current and future researchers who may want to track small changes in environmental parameters, Biosphere 2 had nearly 2,000 sensors. These electronic sensors allowed us to measure variables like carbon dioxide, pH, temperature, and humidity while analytic equipment makes other detailed water and air analyses possible. Each sensor was designed to analyze one variable and its data was then connected to a series of command and relay computer stations that made up the complex nerve system of Biosphere 2. The data they generated was archived at Mission Control in duplicate form for safety. Since these sensors generated data either continuously or on a fifteen-second or three-minute averaging, that made for a lot of data—about 10 megabytes per day!

Some have called Biosphere 2 a 'cyclotron for ecology', and in some ways this is an appropriate analogy. Just as a cyclotron speeds up the motion of atoms to allow physicists to discover new subatomic phenomena, Biosphere 2 speeds up certain ecological cycles, and offers new ways to study the fundamental processes of life on Earth. Carbon dioxide enters and leaves our atmosphere thousands of times more quickly than in Earth's vast atmosphere. Water moves from the ocean, to atmosphere, to drinking water, to raindrops in a period of weeks. We lived by the rhythm of these accelerated cycles, and one calendar year of study in Biosphere 2 allows us to collect better controlled and far more data that we could have in the same year in the Earth's environment. Of course, one interesting difference is that the cyclotron pursued the investigation of the smallest physical particles, and Biosphere 2 the largest laboratory ever built for the study of life systems, at both the micro- and the macro-systems level (including humans).

CARBON BUDGET

Among the most potentially important of the research projects carried out was modeling the carbon cycle. Where was all the carbon when we closed, and how would it change over time? This could provide insights into subtle mechanisms at work in Earth's carbon cycle—now a hot topic of both scientific study and public concern because of the threat of global warming.

Ten percent of the approximately five billion metric tons of carbon added to the Earth's atmosphere each year is unaccounted for—we just don't know where it's going. The oceanographers claim that the carbon is being stored in terrestrial soils and biota, but the forest ecologists say that this is impossible. They argue that because of the increased deforestation worldwide, carbon is being released in forest areas faster than it is being taken up in biomass. They point their fingers back at the oceanographers: the missing carbon has got to be in the ocean waters. And there are some scientists who think that since all the numbers we have are so imprecise, there may not be any missing carbon at all!

Carbon moves through both living and non-living portions of the biosphere (air, water, soils, rocks, and living organisms), but never before have the actual flows, rates, and amounts been measured within the framework of a finite, tightly sealed ecological system. We were able to accurately determine these things in the complex world of Biosphere 2, which included many of the most important components of the global biosphere. Even the rates of movement were not static. They depended on temperatures, light levels, rainfall, humidity, carbon content and molecular structure, nutrient availability, and so on. Carbon models used today by scientists and government officials lack sufficient detailed supporting data and are over-simplified, steady-state diagrams. Therefore, the real carbon situation in the world today is much in dispute.

Knowing the sources and sinks of carbon is critical if, for example, we want to make any predictions about the effects of global warming (caused by increased levels of carbon dioxide, methane, and other greenhouse gases in the atmosphere). Over the past two hundred years or so, levels of carbon dioxide in Earth's atmosphere have risen from about 280 parts per million to 360 parts per million. Most of the increase in carbon dioxide is caused by the burning of fossil fuels, but a certain amount is also attributed to the worldwide destruction of forests and other ecosystems. Still, we know little about the sources of carbon dioxide and even less about where carbon dioxide is deposited or transferred. Without dynamic and accurate models of how carbon cycles work and detailed field studies around the world, even the bare facts of global warming (not to mention the implications for human technology and ecological health) will continue to be unknown. So, appropriately, measuring the effects of the rising carbon dioxide on vegetation and corals was the first of the great opportunities we had to work with in our two-year experiment.

We started off by tackling the question of how much carbon resided in plant tissue. Before closure, ecologists and surveyors mapped the entire Biosphere, identifying the exact location and size of around 10,000 plants. During those two years, we re-measured subsets of these plants to get an idea of how fast they were growing. Using standard techniques and a few simple measurements (the diameter at the base of the plants, the diameter at breast height if tall enough, overall height, and spread of the leaf canopy), we estimated the amount of biomass and carbon each plant contained. From around eleven metric tons of biomass when we closed, Biosphere 2's biomass increased to around twenty metric tons judging from the partial resurvey we conducted in July 1993.

To determine how plant growth in Biosphere 2 differs from 'normal' because of our lowered light, higher carbon dioxide, and lack of strong winds, we would have to re-measure every plant during the transition and perform a limited amount of destructive sampling (that is, some plants would have to be cut and dried to determine carbon content and density). Since many of the first canopy trees had fulfilled their function of providing sunlight protection during the early growth of the rainforest community, they would be cut down during transition and used for this study. These measurements would provide valuable data about how our ecosystems matured, how elevated carbon dioxide affected their growth patterns, and which types of plants dominated their respective ecosystem.

Another key to the model of carbon flows was the soil reservoir which contained over ninety-five percent of the organic carbon in Biosphere 2. Despite its importance in all ecological processes, many of the fundamental processes of soil formation and evolution are still little understood. This was not so surprising given that every gram of soil contains millions of bacteria and fungi which supply a wide diversity of metabolic and ecological functions. But therein lies the power of soils—they are important agents in ensuring that nutrients remain available to other forms of life.

To sustain the wide variety of special habitats inside our world, our soil geologists created more than thirty different soils and marine sediments modeled on Earth's soils. Some of our soils were twelve to sixteen feet deep to support the massive growth of trees over the coming decades. We had a great opportunity to observe the changes that occurred in the soils over time. To do this, a complete set of soil samples was taken before closure, and then the soils were resurveyed at periodic intervals. Chemical analyses, in particular analyses of carbon, nitrogen, and phosphorus, showed us how these soils were changing. And from this, we began to understand the different vegetation groups these soils supported, and how our differing seasonal patterns influenced them.

When news of Biosphere 2's elevated carbon dioxide reached researchers in the United Kingdom, William Chaloner, the head of England's International Geosphere/Biosphere Program, and David Beerling of Sussex University had reason to be especially interested. They were paleobotanists and had developed techniques for studying how plants in Earth's early history responded to carbon dioxide levels. They studied ancient plant leaves preserved in fossils and calculated the density of stomata (openings in the leaf for air and water exchange). Stomata density can then be correlated with geologists' estimates of the concentrations of carbon dioxide in Earth's atmosphere millions of years ago. Biosphere 2, they saw, gave them a chance to check their work by comparing the stomatal density of a wide variety of living plants exposed to high levels and extreme fluctuations in carbon dioxide that only occurred on Earth over a span of hundreds of millions of years.

To support these studies, we collected plant samples, exported them out of Biosphere 2, and shipped them by express mail to laboratories in England. These studies provided information about some of the plant responses we may expect if global carbon dioxide continues to rise.

This was only the beginning, because rocks, water, and animals also contribute to the flows and fluxes of carbon. Therefore, we monitored the various aspects of the environment until all possible sources and sinks of carbon had been meticulously tracked. Because we had a closed environment and a great deal of life activity, we could track carbon dynamics because plants use some naturally occurring isotopes of carbon preferentially. From these data and the continuous monitoring of our carbon-containing gases, we would create a dynamic computer model of Biosphere 2's carbon cycle.

WHERE IS THE OXYGEN GOING?

News of the oxygen decline attracted Dr. Wallace Broecker and one of his graduate students, Jeff Severinghaus, at the Lamont-Doherty Earth Observatory of Columbia University. Dr. Broecker agreed with our initial assessment that the oxygen was probably being lost in the soils, although the ocean and even the concrete basement and rock work were also suspect. But which soils and by what mechanisms? He targeted one soil type, Wilson's Pond soil, which we introduced into the agriculture and in a few places in the wilderness. The soil was taken from a pond just north of Biosphere 2 that was frequented by cattle and therefore was rich in organic material. Broecker theorized that it was also probably largely anaerobic (without free oxygen), since part of the year it was flooded by monsoonal rains, and in its new environment in Biosphere 2 would be an ideal sink for oxygen. There were many other possible mechanisms that could be at work as well: the oxidation of soils containing iron or sulfur compounds (when a soil becomes oxidized it absorbs oxygen); the denitrification of nitrogen; and the formation of caliche (calcium carbonate minerals) in certain of our soils. With some 30,000 tons of soils inside Biosphere 2 comprised of about thirty different soil types, it would definitely take some detective work to track down the culprits.

Calcium hydroxide is a major component of concrete; the same chemical which formed a key part of our chemical recycling system for carbon dioxide. When calcium hydroxide reacts with CO_2 calcium carbonate results (a powdery form of limestone). It is known that CO_2 reacts mildly with concrete, penetrating it shallowly. But in Biosphere 2 with its elevated levels of CO_2, its vast surface area of simulated rock cliffs and mountains, its concrete floors and structural pillars, could this mechanism be a significant player in the oxygen mystery? And if it is, what ramifications does concrete have for oxygen levels in our Earth's atmosphere? To address this question, twelve one-foot-long cores were taken from concrete walls and rocks throughout the Biosphere and of concrete poured at the same time outside Biosphere 2 to analyze their composition. And indeed, when these analyses came back, our mystery was largely solved. In the higher CO_2 environment of Biosphere 2, about ten times as much CO_2 was being absorbed by the concrete than in Earth's atmosphere. So the oxygen was being tied up by first oxidizing some of the organic matter in our soils, which was then released as CO_2 that was being absorbed by the concrete. Sophisticated carbon isotope analyses showed the signature of the CO_2 in the concrete was a perfect match for what we were looking for.

There is an interesting lesson to be learned from this: even in a small man-made biosphere, the complexity of interactions combined with our ignorance of basic information about how ecosystems and biospheres function made surprises and discoveries inevitable. In a sense, noticing then conducting detailed research on the unpredicted oxygen depletion and finding its surprising cause were the kinds of discoveries the Biosphere was built for.

As a spin-off of the oxygen question, Wally Broecker introduced us to two scientists, Dr. Martin Wahlen, Professor of Geochemistry at the Scripps Institute of Oceanography, and Dr. Mike Bender, Professor of Geochemistry at the University of Rhode Island, who were both interested in monitoring both oxygen and carbon isotopes. An isotope of an atom is a form of a chemical element that differs from the regular element in its atomic mass. For example, the most common form of oxygen has eight protons and eight neutrons, giving it an atomic weight of sixteen. One of its isotopes, O_{18}, has eight protons, but ten neutrons, making it slightly heavier. Isotopes of carbon and oxygen are present in Earth's atmosphere, water, and biomass at specific and fairly constant ratios. Will they be found with the same ratios in Biosphere 2? How do they behave? Gaie and Mark began collecting fresh growth of plants of different types from around Biosphere 2 every month for analysis of carbon and oxygen ratios. In addition, periodically air samples were exported for analysis for these studies. Our preliminary findings suggest that the amount of O_{18} in Biosphere 2 air was higher, due to the action of photosynthesis and plant and animal respiration. In addition, patterns of how different types of plants are favoring one isotope over another emerged. Through the use of isotope analysis, we were able not only to identify the mechanisms involved, but also date when the reactions occurred because Biosphere 2's carbon and oxygen isotope ratios were different from Earth's and steadily changed over time.

One media article published about Biosphere 2 stated that the loss of oxygen was due to the fact that humans were consuming it all—an incorrect idea, since our oxygen decline is on the order of 1000 pounds of oxygen every month. Eight human beings couldn't possibly consume this much oxygen. Indeed, all the oxygen humans consume is released as an equal amount of carbon dioxide when we breathe. This carbon dioxide is then used by plant and algae photosynthesis, releasing the same amount of oxygen. Atmospheric oxygen and carbon dioxide are inextricably connected.

Some of us never formally studied chemistry. But for the first time in our lives, the world of atoms and molecules had become an essential part of our thinking. Chemistry was no longer just an arcane discipline studied in textbooks or laboratories. It was integrally linked to our well-being. Finally, in August 1993, the brilliant detective work of Jeff Severinghaus, Wally Broecker, Bill Dempster, Martin Wahlen, and Mike Bender led Jeff to the discovery that most of the oxygen was being trapped as calcium carbonate in the unsealed concrete that protected the stainless steel liner around the floor and basement walls of Biosphere 2. This makes us wonder about the effects of increased carbon dioxide on Earth interacting with our expensive concrete infrastructures.

OCEAN RESEARCH

To study the ocean, a team of marine researchers including biogeochemists, microbiologists, ecologists, and coral experts worked with Gaie in tracking the health of the largest man-made coral reef system ever created. Coral reefs have suffered greatly and are under enormous pressure from human activities. Because the mechanisms responsible for their ill health are far harder to determine, the detailed environmental studies of water quality, light, and temperature which were possible in Biosphere 2's coral reef can be especially useful in determining what factors are critical to coral vitality.

The exuberant Dr. Phil Dustan from the College of Charleston, an authority on coral reefs, joined the marine team in 1992. Phil spent years studying the Florida reefs, noting their decades-long deterioration primarily due to human impacts. He was excited by the opportunity to monitor the health of a carefully measured system of corals through the two-year experiment and beyond. Corals are the key indicator species of the reef's health, and their preservation in Biosphere 2 was one of our major challenges, just as in Earth's oceans.

Beginning in July 1992, Gaie began to take quarterly surveys of the reef using an underwater video. The tapes were sent to Phil who developed an effective system of monitoring the health of corals using a video program. Humans can be quite subjective in their assessment of how corals look, but using his video system, we were able to send Phil images of the ocean from before and after closure to track all the coral colonies. Using a color stick and ruler against each image, it was possible to quantify the health and vitality of the corals by monitoring minute changes in the color of their tissue as well as observations about their overall health. For Gaie or Laser working alone to have done such rigorous surveys and recording of information would have been impossible given our work load, but Phil and his graduate students made these observations from video tapes.

Phil Dustan's participation was particularly timely, because it was in March 1992 when Gaie noticed the first signs of bleaching and a possible coral disease: a white band encroaching the outer rim of some of the brain corals. Corals had mixed success since their introduction into the Biosphere 2 ocean almost two years earlier. Some thrived, some died, some struggled. Bleaching occurs when the algae that live inside the tissue of a coral die. (It is the algae that give coral its colors.) It is possible, however, for algae to re-inhabit the tissue at a later stage and return color to the corals. There seemed to be no particular pattern why some individuals were in excellent condition and others were not. This was a vital question that we wanted to track in more detail.

We sailed through the first six months without any problems with the corals. We anticipated that winter would be the most difficult time for the reef because of low light and high carbon dioxide levels, so were surprised that springtime sparked this new shift. The advent of bleaching and the coral disease coincided with increased light levels and raised water temperatures (from seventy-six to eighty degrees). Phil was able to confirm the presence of a bacterial disease called, appropriately, 'white band disease' that he commonly observed in Florida. Somehow, the bacteria must have been lying dormant ever since we introduced the corals. With time, the bacteria became destructively active.

Later that spring, ocean conditions began to shift again—this time to an over-abundance of algae: there was a small chlorella algae (we called it the 'grey slime') that covered everything, bubble green algae that grew around the hard corals, red spiny algae that grew among the soft corals, and a calcareous red algae that grew along the lagoon floor and reef rock. These four algae species in particular had become terrible pests and continued as significant weed problems throughout the two-year experiment. These algae are quite different from the tiny algae that actually cohabit the coral tissue. These macro-algae, if left alone, would inevitably cover the coral tissue and prevent the corals' algae from receiving enough sunlight, and the coral polyps from ingesting the plankton it feeds upon. Like Linda and Mark in the terrestrial ecosystems, Gaie had to periodically weed the ocean system in her role as keystone predator.

We suspected that the shifting light levels from season to season could be a fundamental cause of coral stress. Corals are not used to such changes, and it must be particularly difficult once winter gives way to the bright sunlight of spring. To further address this question, Phil sent us a light sensor to compare with the one we permanently mounted underwater. He also conducted some comparison studies with the reefs of the Yucatan of Mexico, where most of the corals had been collected, and along the reefs of Belize. Indeed, he found that light levels in late spring may be too bright for many of the corals that have done well in the winter. We then began to consider if our reef should have its own distinctive pattern: a migrating coral colony that moved up the reef in the winter and down the reef in the spring. We did not follow such a strategy; instead we focused on the necessity to increase the quality of the water using the skimmers and macro-algae weeding, as well as man-agement of grey slime. Together, these simple coral reef practices were important tasks that seemed to take care of most of the problems as we continued to track the corals over the next year. In March 1993, there was another slight sign of stress with the arrival of spring light levels, although nothing like what had been experienced the year before. This time around, we were also mindful of keeping water tempera-tures on the cool side. Even one degree makes a difference.

Gaie and her colleagues started a coral reef research program at Turneffe Atoll in Belize to study a natural reef in conjunction with our artificial reef to compare vectors like the health of coral species, the chemistry of the water, and the diversity of the microbial community which forms the base of the ecosystem. Dr. Donald

Spoon conducted a preliminary study in January 1992 where he found eighty-two species of microbiota in the unpolluted lagoon of Calabash Island in Belize as compared to seventy-five species in the lagoon of Biosphere 2. What this tentatively indicated was that the microbial population of our little lagoon remained remarkably diverse even though it was far smaller than the unspoiled wild lagoon in Belize. For over two years, the population of microbiota had not diminished inside the artificial system even though it was subjected to mechanical pump action.

Phil Dustan and Dr. Judy Lang, a coral biologist at the Texas Memorial Museum who joined the marine team in the Spring of 1992 as well, were particularly interested in tracking individual corals over time. Phil started a baseline study of the corals in Belize to set the initial parameters. Their past years of research in the Bahamas showed that individual coral colonies of the same species have different responses to the environment and that the ability of corals to thrive in Biosphere 2 or in the Bahamas seemed to be specific to the individual and not to the species. This was not obvious, in fact, even counter-intuitive, when we started the two-year experiment. One consultant to the marine systems advised us that we would lose at least thirty percent of our species; others suggested even greater losses. But as of the summer of 1993, Phil was still tracking over 500 colonies of corals via the video surveys. Instead of vast species extinctions, we found that individual variation is critical. Some of the colonies continued to flourish while others showed a slow reduction of tissue. It was this individual response that we wanted to understand in more detail, and which formed a basis for future research.

Dr. Bob Howarth and Dr. Roxanne Marino were a delightful team from Cornell University working in marine ecology and chemistry. They managed to bridge the worlds of the pro-active environmentalist groups and the academic university systems. Bob was well-known for speaking on behalf of Greenpeace environmental crusades with indisputable scientific fact at his command. He was one of the world experts in the impacts of oil pollution on marine systems and editor of the journal, *Biogeochemistry*, his other area of expertise. Roxanne helped Bob run the laboratory at Cornell and later earned her PhD and joined the faculty. Together they organized a program for tracking the marine chemistry dynamics and biogeochemical cycling of nitrogen and phosphorus in Biosphere 2, which would be developed to include similar observations at the pristine Turneffe Atoll reef community of Belize.

Of course, all marine research projects cannot really be separated from one another. Each specialty is an aspect of a total system, not an independent reality in itself. To reverse harmful trends, we needed to learn more about the precise mechanisms which lead to vitality or disease in corals, and then develop appropriate means of intervening to restore the ocean's health.

LEAF LITTER AND DECOMPOSITION

Matt Finn, a doctoral student at Georgetown University and research ecologist for SBV, worked with Gaie and Mark to collect data for his dissertation on the comparison of the ecology of the natural Florida Everglades marsh system with its replicate marsh in Biosphere 2. The Biosphere 2 marsh was designed to represent an area that spans over 100 miles from freshwater inland marshlands to the offshore red mangroves, and indeed our relatively tiny 4,200-square-foot marsh was extremely diverse in its community structure. It was packed with species which have not only been maintained but have grown several times their height since their introduction. This system was as complex and diverse as what is found in the real wilderness.

One of the key studies initiated in the marsh, and then expanded to include all the terrestrial wilderness biomes, was that of leaf litter and decomposition. Litter, to an ecologist, is quite different from the unwanted junk clogging up your gutters. It's the fine rain of leaves, twigs, flowers, and fruit that plant communities drop on their soil during the year. The accumulation of litter is an important way the soil is nourished, replacing some of what the plants take out. About 130 square, mesh litter 'baskets' sat on short plastic legs in our rainforest, savannah, thorn scrub, desert, and marsh patiently collecting the material that rains down from above. At the end of each month, Mark, our resident litterbug, moved on hands and knees with a small whisk broom and dustpan collecting the material from each litter basket. In the marsh, the litter baskets sat on taller legs, above water level, and the litter was collected while wading through the water. In this ecosystem, litter fell into the water and floated until it slowly decomposed and sank to the bottom sediment.

These bags of litter, one for each of the 130 sites, were sent out through the airlock with the monthly exports to be dried and weighed in the outside labs and then chemically analyzed. From these studies, we got a microscopic view of the topsoil—the interface between the life below and above. They also showed us how each biome differed in its nutrient cycling. The numbers reflected either the large tropical leaves and branch prunings of the rainforest, or the scattered litter islands found under desert vegetation.

We also set up a decomposition study to complement the litter study. In each ecosystem, mesh bags were made with identical weights of leaves from the dominant plants of that system. These were sealed and placed on the top soil or sediment on September 10, 1992. There were twelve such bags made for each area so that each month for one year, one of the bags is collected and analyzed. This gives us the ability to determine how fast the material disappears and is incorporated into the soil. In the marsh these bags were placed in the system and tied down with fishing-line so that they wouldn't float away.

Matt was conducting an exact parallel study in the Everglades with Dr. Pat Kangas of the University of Maryland. Over the coming years, he will study nutrient circulation in our young and growing marsh compared with its parent marsh, which is subject to hurricanes, strong tides, human pollution, and theft of water. The Everglades may be the first U.S. national park facing the danger of dying as an ecosystem. It is threatened because farm irrigation diverts the fresh water that normally flows through and replenishes the marshes. Its other enemies are the chemical residues that drain through it from fertilizers, herbicides, and insecticides now affecting the delicate aquatic food webs.

WILDERNESS STUDIES

One of the most interesting things about Biosphere 2 was studying the dynamics of relatively small populations of plants and animals. This was invaluable data because it paralleled concerns about biodiversity in Earth's many ecosystems where human impacts, such as road building and forest clearing, have created smaller habitat islands where smaller populations of organisms struggle to survive. Among the problems this creates are genetic 'bottlenecks', when smaller populations descend from a handful of ancestors. Some baseline studies of genetic diversity in insect, fish, and plant species were conducted before closure and would be repeated at re-entry to determine whether such inbreeding caused problems.

To further these studies, we collected a sampling of insects by setting out jars with rotting fruit as bait. The insects were freeze-dried in the medical laboratory inside Biosphere 2 for later examination by experts. We collected leaves from a few individual plants of several species in each biome for genetic analysis before closure and repeat examinations would be done during transition. Additionally, two species of a fresh water fish, (*Gambusia affinis and Gambusa holbrooki*), were stocked before closure as a special population study with the University of Georgia under supervision by geneticist Dr. John Avise. A few were trapped at the one-year anniversary and frozen to study the effects of competition and hybridization.

To track insect changes since closure, we used several techniques to capture and export individuals for identification by Dr. Scott Miller of the Bishop Museum in Hawaii, who had been working with the project from its first design stages. Every month, we opened pitfall traps (deep plastic containers dug in so that their lip is at soil level) for twenty-four hours. Insects that fell in were collected, preserved in alcohol, and sent out for identification. For night-flying insects, we hung black lights in front of a piece of white cloth and collected the insects attracted. To monitor insects which resided in leaf litter, we used a special device called a Burlesi funnel where a low-wattage light creates heat to drive the insects down through the mesh of a funnel and to the bottom of the bucket where they are preserved in alcohol.

To check on how our plants responded to their somewhat unusual environment, we also performed monthly checks on their flowering, seeding, and overall growth. Linda collected detailed data in the rainforest, savannah, thorn scrub, and desert, and Mark sampled the marsh, beach, and the salt-loving plants that grew in the halophyte garden on top of the wave generator. During these repeated observations, we noted insects and animals, and made other observations as well. One long-term research activity was correlating the phenology of the plants with trace gas dynamics in our atmosphere, as plants release a variety of minute gases during their growth and flowering. To study what was released into the atmosphere, we collected monthly samples of pollen which were sent to Dr. Mary K. O'Rourke at the Respiratory Science Center of the University of Arizona for analysis.

It was a common sight to see SBV's desert consultant, Dr. Tony Burgess of Tucson's Desert Laboratory, wandering around the outside of the Biosphere as he made his monthly observations, which were then compared to the ones we made inside. Tony was a self-declared desert rat, a colorful character with a distinctive bushy red beard. He was extremely knowledgeable about the ecology of the desert and thorn scrub ecosystems and carefully tracked the evolving structure of these biomes.

SENSORY STUDIES

Our two years inside gave us the opportunity for some unusual medical research. Working with Dr. Gary Beauchamp, Director of the Monell Institute, a leading institute in human sensory research, Roy initiated experiments to observe changes in our sense of smell. By using 'smell' bottles which contained a variety of concentrations of two distinctive smells, roses and spoiled milk, they attempted to ascertain if we had developed a reduced or enhanced smell capacity. To do this, Roy gave each crew member a whiff of a bottle with a smell/odor and one without any fragrance to test our odor sensitivity.

To monitor our overall level of stress, Roy collected a series of daily urine samples from each of us which was examined for the hormones associated with stress. Urine was also collected early in our adaptation to the calorie-restricted diet and sent out for analysis to ensure that we were receiving adequate amounts of protein and other nutrients.

Another intriguing study is being conducted by Dr. John Laseter, Director of Accu-Chem Laboratories in Texas. Laseter was looking at what environmental compounds were present in our bloodstreams. During the two-year closure, periodic blood samples were sent out for analysis. One early finding of interest showed that levels of pesticides and other toxic chemicals increased in some of the biospherians' blood after we started living in Biosphere 2. The reason for this was that we lost considerable weight in the initial adaptation to our diet, and many of these compounds which are normally tied up in fatty tissue were released into our bloodstreams.

RESEARCH WITH THE RUSSIANS

There was great interest in Biosphere 2 from the two Russian institutes who had led their country's efforts in developing biological life support systems for space. The first of these was the Institute of Biomedical Problems in Moscow, where Yevgeny Shepelev became the first human to live with biological life support in 1961, when he spent twenty-four hours in a chamber where chlorella algae regenerated his air and purified his water. At the Institute of Biophysics in Krasnoyarsk, Siberia, these algae-based systems were further developed with the Bios-3 experiments in the 1970s and 1980s, where they achieved six-month closures with a dozen food crops supplying half the food and providing nearly all the air and water regeneration for crews of two and three people. Both institutes had participated in a series of international workshops in closed ecological systems that SBV and the Institute of Ecotechnics organized since an initial meeting at the Royal Society, London in 1987 to pull together all the researchers working in this area.

In addition, we did collaborative research with both Russian institutes. Scientists from Siberia worked with us during our pre-closure investigations of the health and medical aspects of closed ecological systems. Dr. Lydia Somova, their top specialist in microbes essential to human health, did a survey of the biospherians to see if their microbial populations simplified during closure as occurred in the Russian experiments. With the Institute of Biomedical Problems, we participated in the first space flight study of a functioning, small aquatic ecosystem containing guppies, chlorella, and bacteria. It flew on a Biosatellite and spent twenty-one days in weightlessness in September 1989. In comparison with the closed aquatic systems on the ground set up for a control, the ecosystem functioned perfectly, and all the components of the system reproduced themselves successfully. During our two-year closure, we sent samples of algae growing in varying environments of Biosphere 2 to IBMP in Moscow for identification. We worked together to evaluate if culturing some of the more productive of these in order to release free oxygen may help to counteract future oxygen depletion in Biosphere 2. The results were also applicable to understanding other smaller life support systems for use in space.

RESTORATION ECOLOGY

Biosphere 2 had to be a pioneer in restoration ecology engineering, since all the biomes had to be built from scratch. The reef, for example, started out as a giant rectangular steel tub that came to operate as a self-sustaining coral reef community isolated in a new little world, which put us in a unique position to understand what makes reefs tick. Like rainforests, reefs are endangered around the planet, but we are not sure exactly why. We have some hints: we know that pollution, the dumping of wastes, and the physical impact of boats and divers are all harmful. The overriding question is how to reverse the trend of ocean degradation. How can we restore these damaged areas? What is required to restore the system so that it will resist further human impact and continue to flourish? Restoration ecology is a new field that attempts to answer these questions.

Some restoration ecologists are purists who maintain that the ecosystem must be returned to its original natural form. They feel that not only must the diversity of species and complexity of the community structure be protected, but that people should stay out completely. Given our uncertainty about whether human contact can be made non-destructive, this approach is understandable. At the other end of the spectrum, restoration ecology can be looked at as a science that discovers what the function of an ecosystem is with regard to global or local ecology and then recreates that system to address this need. This is perhaps a more practical approach, one which tries to ensure that ecosystems are able to evolve and adapt to changing environmental conditions.

Our experience in creating the diversity of ecosystems inside Biosphere 2, and then intensively studying how they developed and matured, gave us powerful tools to apply to restoration ecology in damaged natural systems. Our approach sought to unite human intelligence with appropriate technologies that assist rather than degrade the environment. In the twenty-first century, we can hardly attempt to recreate a world without humans or technology, but we can utilize our strengths to begin to restore what we have degraded.

A NEW RELATIONSHIP WITH THE BIOSPHERE

Bill Jordan of the University of Wisconsin Arboretum, a pioneer in restoration ecology, once asked us, "If you can measure anything you want and you know everything about the physical nature of your little world, doesn't it leave you without a feeling for the unknown? Don't you miss going into the wilderness to lose yourself in its beauty and expansiveness? If you can control elements like rain and temperature, what is there left to seek? Do you experience being responsible for your world as a burden, as a constraint on your freedom?"

Our answer is that, quite to the contrary, Biosphere 2 was a cathedral of life; full of different ecosystems abundant in beauty and diversity that is fulfilling. Our integral connection with it as we worked alongside our plant and animal communities, gave us great joy and courage for how we all need to be especially in these times of frightening planetary change. For the first time, we had a sense of our place in a world where we could very quickly and clearly see the results of our actions. It is possible to be a positive force in the synergy of life, a necessary element contributing to, as well as benefiting from, the overall health of the system.

We led, organized, and participated in scientific conferences around the world, and consulted with our Mission Control via video conference.

CHAPTER 11

COMMUNICATION THROUGH THE ELECTRONS

"What prevented past ecosystem ecology from becoming an experimental as well as observational science was lack of vision about ways of achieving large-scale closure and unwillingness to face the effort and capital that it required… somehow the necessary vision, commitment, and capital came together here and ecosystem study as an experimental reality bas been born."

– **Professor Harold Morowitz**, George Mason University,
Re-entry day talk, September 26, 1993

THE PAPERLESS OFFICE

OFFICES ALL OVER THE WORLD are trying to use computers to minimize the use of paper, but few are succeeding. In fact, in most offices, the use of computers has probably increased the consumption of paper. However, to reduce consumables and make a recyclable world, SBV decided on an all-out effort to create a paperless office.

The offices containing our computer equipment were located in the command room, a semi-circular domed room on the second floor of the habitat. Each crew member had a desk with a computer equipped with a modem. We had a conference table with the ability to link up with other video sources, and some of the larger computers that could call up all the data from the sensor systems of Biosphere 2 were also housed here. Biosphere 2 had miles of computer, video, and phone cables linking it to Mission Control and the rest of the outside world. So, despite physically leaving the world, we were determined to continue operating in it informationally by learning to use these electronic communications.

The whole staff of Space Biospheres Ventures, including the biospherians, and some of our research consultants in the United States and Europe were connected by email, then a quite new way to communicate. All of the memos and reports generated inside Biosphere 2 are sent via computer to the outside where they are read, printed for distribution, or faxed to the appropriate people. In general, SBV stressed the need to minimize paper use, so employees were encouraged to do as much of their business as possible on the computer.

Inside Biosphere 2, we used virtually no paper at all. Since we could neither make it nor re-use it, paper had no place in our world—especially since we would have required tons of it to meet the demands of our communications. It was clear right from the start that our information had to be transmitted and exchanged 'through the electrons'.

The electronic medium made the most sense both for closed systems like ours and for communication in space exploration. Even in the Earth's biosphere, the production of paper makes heavy demands on the environment. Some of the objectives we had in creating a paperless office were related to how we foresaw life in the twenty-first century. Our plans exceeded what was available back in 1991. For example, when we came inside in September, there were no widely available technologies for transmitting faxes without paper. Several months later, the technology to receive faxes to our computer came on the market and was incorporated into our systems.

The tempo of our lives, combining sustainable agriculture, naturalist observations, technical operations, and a worldwide network of colleagues, would be impossible without telephones and computers. Biospherians received mail transmitted via email and fax. But the various paper-based communications addressed to us were dealt with on the outside by some invaluable assistants who screened our business and personal mail. Unfortunately, junk mail continued to arrive for all of us. Luckily, SBV had a paper recycling system so that this mail could be chucked into recycling bins. Each of us designated someone on the outside to handle personal mail. This is either read to us over the phone, brought to the window, or most commonly, sent in via fax. If it came through to our computers, we could file it in electronic folders and save it for later reference or reply.

Sometimes, after one of us had done a communications linkup over telephone or video with a school class, thirty or forty letters from the students would come in over our fax system with further questions, their thoughts about becoming biospherians, or even suggestions and recipes for us to try.

Each biospherian was given bound notebooks to use as diaries or journals. We also used them to record scientific data and observations, since some of us still preferred to make our notes by hand in the old-fashioned way. However, with the exception of scientific research papers, books, and the odd 'post it' pad for note taking, there was virtually no loose paper inside Biosphere 2. When we first started to operate without paper, it was extremely challenging. But it had become a part of our daily life. Some of us had portable laptop computers in our apartments. Filing and handling all the hundreds of faxes, reports, memorandums, and other correspondence on a computer took far less time than dealing with all those piles of paper that accumulate on a desk.

Just because we lived in a separate world does not mean we got away from the responsibilities of normal citizens on the outside. We had newspaper articles faxed in, or on the computer, but we couldn't read the newspaper in the ordinary way. We watched the news on TV and listened to the radio. Some of the biospherians expressed disappointment that they no longer could see the comics in the newspaper. This became of interest to all eight of us when Biosphere 2 began to be featured in that section. Friends taped the comics to a window of the habitat or faxed them in for us to see. One of our favorites showed up at Christmas. Santa Claus landed with his reindeer on the Biosphere 2 space frame roof and is rapping on the space frame structure, trying to get in, with a big question mark hovering over his head!

When we entered the Biosphere, the Soviet Union was one country and apartheid was still legal in South Africa. Nothing else compares in significance to these momentous changes, but there was a good deal of news pertinent to our project. For example, we followed the Environmental Summit in Rio de Janeiro, Brazil with great interest. Especially poignant to us is coverage of environmental disasters such as the large Exxon Alaska oil spills and the oil wells set on fire in Kuwait, continued pollution and degradation of the Earth's ecosystems, and the arguments back and forth on who is responsible for this or that environmental crisis. Inside Biosphere 2, we knew who was responsible for our world and pointing fingers at wrong-doers was less the issue than taking action to help.

We followed the presidential elections closely. In the fall of 1992, the four of us who are registered voters in Arizona were able to vote in the primary and in the November county, state, and national elections via absentee ballots. Two voting officials from Florence, Arizona (the county seat) came down with the ballots and held them up at the meeting window. Each biospherian voted in turn, reading out over the phone the candidates they wanted to vote for. One of the officials signed on our behalf and then turned in the ballots.

Paying taxes was another responsibility we couldn't escape. Since we all remained either citizens or residents of the United States, Uncle Sam required us to pay taxes on our salaries and other income. To make this easier, we each gave someone on the outside power of attorney to sign official documents for us. Tax forms were electronically faxed to us and after we work them out with our accountant or friends, they go out to be signed on our behalf. Then, whomever we had authorized wrote checks to the IRS and sent them in. Any refund or salary checks were deposited by our outside helper in our bank accounts. Since our opportunities for spending money were more limited than usual (we had free room and board), many of us had piled up savings like never before. Other mundane matters of life were also handled electronically with little difficulty. These included things like renewing expired driver licenses, making investment and banking decisions, wiring money, and paying bills.

MANAGING BUSINESS AT A DISTANCE

In the modern world, businesses are increasingly run by people scattered all over the world. We biospherians had also been able to use the communications grid that covers Earth to start new companies, have board of director meetings, and even arrange and participate in scientific workshops and conferences. Why go through all that jet-lag, culture-shock, lost baggage, visa applications, passport and bomb checks, canceled flights, foreign money exchanges, haggling with taxi drivers, hotel reservations, airline food, and other wear and tear on the body and mind when all you want to do is organize and discuss business?

A perfect example is the non-profit foundation called the Planetary Coral Reef Foundation (PCRF) that Gaie and John Allen started shortly before closure in order to compare the ocean reef in Biosphere 2 to those found in natural areas of the Caribbean where the Biosphere 2 reef was collected. The company addressed questions regarding the demise of reefs worldwide, developing techniques to further the restoration ecology of reefs and marshes, and organized studies of these marine ecosystems in relation to global ecology. The field site chosen was Belize because Blackbird Caye Ltd., a Belizean company, offered to cooperate with PCRF in a joint venture: they'd organize an eco-hotel for reef and marsh education and PCRF would set up a research center alongside some of the most beautiful reefs left in the world.

Since the PCRF was only just formed by the time Gaie walked through the airlock door in September 1991, she still had to organize the legal process necessary for establishing a non-profit foundation with Michael McNulty of Brown and Bain P.A. She also coordinated the team of scientists who participated in the research program, secured donations and income to support the research vessel *Heraclitus*, stationed in Belize; additionally, she produced a brochure, and conducted meetings with the scientists and board of directors to further the objectives of the foundation. Six months later, she and her colleagues convinced the Belize Government to make

Calabash Caye, south of Blackbird Caye, a national reserve. PCRF was then leased two acres of land to conduct observations of wild, untouched reefs and marsh lands.

In January of 1992, Laser and a friend started another company, Blue Planet Divers, as a subsidiary of PCRF. It provided diving support for the ecotourism business and the foundation's research activities. Laser was also instrumental in establishing a diving school in Belize by purchasing the needed equipment, consulting with the Belize team on business decisions, and encouraging groups, both diving and ecological, to utilize the facilities. All of this was done through computer and phone within Biosphere 2.

Mark also conducts business over the electrons. He was the chairman and CEO of a small, non-salaried, ecological think tank and research group, the Institute of Ecotechnics. Registered in the United Kingdom, it provides consultation services to a number of innovative ecological ventures around the world, including Biosphere 2. Mark convened and ran two annual meetings of the company and its board of directors while ensconced in front of a computer inside Biosphere 2. Since these meetings of the mind take place through the electronic mail of the twelve directors, there was actually no telling where they might be. He suspected from what he knew of their far-flung activities that they may have participated from computers in England, France, Australia, Puerto Rico, Ft. Worth, Santa Fe, as well as in several offices a stone's throw from his office window. Perhaps one of them was traveling and sending messages during a layover in Singapore or Hong Kong, or while waiting for the fog to lift at O'Hare Airport in Chicago.

While this technology is efficient, some may lament that they are unable to read body-language in business meetings held in this fashion. You cannot view the subtle twitching of facial muscle, or the blink of an eye. You are unable to sense eagerness or diffidence, willingness or bluff. On the other hand, conducting business electronically helps to eliminate distractions. Some people are able to communicate better when they don't have a lot of background noise to deal with. Mark was pleased that no one involved in his meetings could hear the noise he was making as he chewed on sugar cane and cracked open roasted peanuts during a break-time snack. Nor could they comment on his messy attire if he was still sporting a couple of mud splats from rice harvesting. All they knew was the text that appeared on the computer screen.

Roy Walford continued to oversee the management of the laboratory he ran in UCLA, which employed about ten people. He was able to put together a lengthy grant application to the National Institutes of Health, with portions written at his lab at UCLA and letters of support from the Metabolic Lab in Phoenix and by researchers at the University of Arizona Medical School.

Taber and Jane even managed to have a house built, which they would move into once they left the Biosphere. They purchased the land, hired the architect, approved designs, and contracted for the building work. While much of this was done through the electrons, it did entail the architect coming by to hold the sketches up to the window for them to see.

SCIENTIFIC WORKSHOPS AND CONFERENCES

Perhaps the most complicated bit of business conducted from inside Biosphere 2 was the arrangement of the Third International Workshop on Closed Ecological Systems, organized by Mark on behalf of the Institute of Ecotechnics, hosted by Space Biospheres Ventures and held on the Biosphere 2 site in April 1992. This was a sequel to the first two International Workshops on Closed Systems. The first was sponsored by SBV and the Institute of Ecotechnics at the Royal Society in London; the second, held in Moscow and Siberia, was co-sponsored by the Russian Institute of Biophysics. This was, however, the first 'interbiospheric' workshop on closed ecological systems—the participants came from two different biospheres. It was also memorable in that many of the pioneers in the field of closed systems, both from the United States and Russia, participated.

Using the telephone and fax, Mark lined up about thirty scientists from universities and space programs in the U.S., Europe, Russia, and Japan to deliver talks during the four-day event. He even chaired part of the meeting via SBV's two-way video system and presented a paper. In addition, all the biospherians were able to participate. During presentations, one video camera took close-ups of slides and graphs and relayed the shots into the Biosphere. Another camera focused on the speaker or questioner from the audience. Besides Mark, Roy and Linda gave talks at the workshop by coordinating with someone in the main conference room who presented their slides and graphs.

Five other scientific workshops on marine systems were organized by Gaie. They involved members of the marine consulting team who were either present on site or convened over the telephone. Sally organized an agriculture workshop and Mark a soils workshop, both of which hosted a number of scientists at Biosphere 2 for meetings spanning several days.

Visiting scientists such as Larry Slobodkin, Richard Evans Schultes, Pat Kangas, and a dozen more gave talks in a guest lecture series which was piped in by video to the biospherians. On other occasions, we ourselves gave scientific lectures and participated in conferences via electronic links. Some of these included talks over a video telephone to the International Space University in Nagoya, Japan; the EcoHarmony conference in Monterey, California; two successive annual black-tie Explorer's Club banquets in New York; and the Case for Mars V Conference in Boulder, Colorado.

The 1992 Annual Explorers Club celebration was a particularly outstanding event. The marine scientist, Sylvia Earle, noted that the meeting was held when explorers now had three worlds to roam, with three separate life support systems. Overhead, orbiting in a space shuttle, was her friend Kathy Sullivan, also a member of the Explorers Club. A second world included the 2,000 explorers of planet Earth who had gathered in New York for the meeting. And in the third world were Gaie, Laser and Mark attending the evening and representing the eight biospherians, explorers of planet Earth's first biospheric offspring.

EDUCATIONAL LINKUPS

Since environmental education was one of the important objectives of Biosphere 2, the crew participated in weekly linkups with schools. These linkups could involve an actual site tour of Biosphere 2 where students interact with biospherians, or they could be conducted through telephone and video. Sometimes these linkups involve conferencing many classes, often from several states, over the telephone at the same time. With use of a speaker phone, all the students were able to ask questions and hear the answers.

The most intriguing linkups have been with classes that designed and built their own 'Biosphere 3' model or mock lunar bases. In one case, thousands of grade-school and high-school students around the US and Canada fabricated a sealed space capsule and lived inside it for three days without coming out. During this time, they considered many of the complex aspects of closed systems such as atmosphere regeneration, human health and physiology, communications, command structure and social interactions, agriculture and aquaculture systems, and recycling. Because people learn best by doing, this project was a great tribute to its organizer and all the educators and students who participated in it.

We had also held a couple of wide-reaching educational links via satellite. One was arranged through the well-known Fairfax, Virginia school district which helped pioneer 'electronic field trips.' These trips introduced students in the US and Canada to exciting research and multicultural encounters. This hookup was seen by about two million students, from grade school to graduate school. On Earth Day 1993, we participated in another major production that was sponsored by IBM. We gave about two and a half million students from Hawaii to New Haven a satellite tour of the Biosphere. Both of these tours featured taped video footage shot in Biosphere 2, live commentary by several crew members, and a question and answer period where students were able to call in directly to the Biosphere and talk with us.

This enthusiastic interest from young people gave us a great boost and enduring hope for the future. Two of the most frequent questions are: "What do I have to do to become a biospherian?" and "What advice can you give us on taking better care of our biosphere?" Though they frequently had a bit of trouble picturing how small Biosphere 2 was—we were often asked about tigers, dolphins, sharks, sometimes even giraffes and whales could be included in their model biosphere—their enthusiasm confirmed our hope that our work would stimulate new ways for people to think about themselves and their environment.

ELECTRONIC CAMPUS

Several of the biospherians had continued their academic careers while inside Biosphere 2. Linda Leigh wrote her application and was accepted to a PhD program with Union Institute of Cincinnati which specializes in innovative graduate programs for multidisciplinary and older students who wish to return to school to complete their doctoral programs. She even took part in her first seminars via phone linkups.

Gaie also prepared her application and was accepted for a PhD program under Dr. Harold Morowitz at George Mason University in Virginia. While she had finished her course work for a doctorate at Yale University before entering Biosphere 2, she had not done her dissertation. She met over PictureTel with her committee from George Mason to discuss her requirements and her proposed thesis to design a biosphere system for a Mars base settlement. Dr. Morowitz was one of the early scientists involved in closed ecological system research as well as being a distinguished thermodynamicist. Fortunately for Gaie, as well as Linda, their acceptance process was simplified by the electrons and neither had to travel across the country for interviews.

Jane and Taber, who had taken advanced technical training with SBV, prepared to enter college programs so they could continue with their interest in space sciences. During our final year in Biosphere 2, they were both accepted into the Honors program at the University of Arizona.

Mark was accepted for a graduate program with the School of Renewable Natural Resources at the University of Arizona. His master's thesis, under faculty advisor Dr. Lloyd Gay, would be a study of an ecological waste recycling system using fast-growing poplar trees. Part of the acceptance requirements involved several prerequisite courses. So during the second year of our closure, Mark signed up for meteorology and geology courses. The textbooks came in on one of our monthly import/exports, and Mark was able to complete most of the course assignments on computer files which were then printed out and mailed by our administrative office.

Mark's final exams were proctored by Barbara Rossi, a representative from the Extended University of the University of Arizona which arranges these correspondence courses. The test was electronically faxed in, and while Mark worked on answering the questions, Barbara kept an eye on him via the video linkup. At the end, he read off his answers, which she filled in on the paper copy of the exam.

Mark's geosciences final involved drawing diagrams, so the exam was imported for Mark to complete under Roy's supervision. It was then sealed in an envelope and sent out on the next scheduled export. Mark did us proud by 'acing' all his courses. For the people at Extended University, these correspondence courses with Mark were a new challenge. A few years earlier, their most far-out students were soldiers stationed in the Middle East during the Gulf War. Now they had a student in another world!

THE FUTURE OF ELECTRONIC PARTICIPATION

Using our electronic facilities, we were able to co-author numerous papers for scientific journals and popular magazines. These were sent out electronically and turned into hard copy for editing, study, and review on the outside.

While we have been pleased by how connected we have been when using the variety of electronic communication devices presently available, we sometimes felt our isolation. Some things that are normally easily arranged take a lot of explaining and haggling to accomplish from a distance—in particular anything that requires a signature. And remember Murphy's Law: what can go wrong, will go wrong. Our electronic equipment was fantastic when it worked, but a migraine when it didn't.

Linda gave us a perfect example of how our isolation could cause problems. A State of Arizona wildlife inspector who read in a newspaper story that we had a curved-bill thrasher inside Biosphere 2 visited us to find out if we had the proper licensing papers to keep it in captivity. It took a lot of explaining to make him understand that the bird 'volunteered' for Biosphere 2 by flying in during construction and had consistently evaded our attempts to export him through the airlock. If filing a paper with him could persuade our feathered friend to willingly depart, we'd get someone with our power of attorney to sign in quadruplicate as many forms as he cared to produce! Our problem was that we couldn't catch him, so how could we get him out? Finally, Linda and the inspector worked out all the necessary paperwork to keep everyone happy.

It had been a challenge to leave the world physically and still play an active part in such a wide array of business, scientific, and educational activities. Our flourishing efforts to create a totally paperless office would be improved when some other technologies were added to our system—larger and easier to read computer screens, for example, and a scanner for converting research papers into computer files that can be read, edited, and stored. The challenge of the truly paperless office has still to be met, but we had gone a long way toward doing so.

Parties and feasts created special occasions to relax and have fun. Here crew members celebrate Gaie's 33rd birthday.

CHAPTER 12

After Hours

"But perhaps Biosphere 2's greatest value has been showing what is possible. How innovative these eight people were in being able to live in a small area and make it work. The real lesson there is not so much a scientific lesson, but showing how people can work together when resources are short, which is going to be the situation in the world."

– **Professor Eugene P. Odum,** University of Georgia,
Boston Globe, February 14, 1994

No going out to the movies. No meals in restaurants. No museums to visit or plays to see. No weekends in the mountains. No summer vacations at the seashore. Entertainment was entirely up to the eight of us. We would have to entertain ourselves or remain un-entertained.

In the early weeks after closure, we scheduled one-hour crews on Sunday. However, it became apparent quite quickly that we needed time off, time for ourselves and our private pursuits, and that Biosphere 2 could operate itself safely for a day, relying on its alarm system for emergencies. We resorted to a formal vote at a crew meeting in early October to clear the decks of any obligatory meetings on Sundays and to take off all the holidays that the staff on the outside did. The vote was unanimous. If we didn't draw the line, work and research could become grimly obsessive, actually leading to a diminution in the quality of the output. Even so, the necessities of Biosphere 2 meant that someone cooked even on days off, others feed animals, and others do the small chores that can't be ignored, like checking water and mechanical systems.

Sunday then became our day of freedom. The only requirement is to show up for breakfast, lunch, and dinner—if you want to eat. Otherwise Sunday is our own time to put into writing, painting, video or music production, rest and relaxation, or strolling through Biosphere 2 to enjoy a swim or walk in the rainforest. Before closure, all of our energy went toward completing the facility and preparing ourselves for the task inside. After closure, it had become essential to find leisure time and use it creatively. We had to make time to play hard as well as work hard!

Some used Sundays to tour the entire Biosphere and catch up with the changes in biomic areas outside our specific responsibilities during the past week. This includes snorkeling in the ocean for the three non-divers: Sally, Linda, and Mark. Four biospherians became avid photographers and five went on from keeping personal journals and scientific observation notebooks to a variety of other writing projects. Gaie, Mark and Sally used the weekends to work on this book; Mark started to write poetry; Sally undertook her Biosphere 2 cookbook; and Laser began an account of his life in Biosphere 2 in his native tongue, Flemish, for a Belgian publisher. Linda worked on a naturalist's account of her two years.

The Biospherian Band, led by Jane and Taber, often held jam sessions on the weekends. We set up the electronic gear on the IAB balcony to take advantage of the acoustic effects of the soaring space frame above. Chirping crickets below added an exotic chorus. Sally played her flute. Jane started to paint. Laser made a video studio in his room, complete with editing facilities and an assortment of cameras. He and Gaie work on interpretive and documentary videos of life inside the Biosphere. Linda became an expert in discoursing and interacting on a computer network called the WELL.

Roy thoroughly enjoyed his special art projects. One of them involved a montage video documentary which he'd put together during closure. He aimed to create a video art piece that could be displayed on nine separate screens. He also did photography work where he tried to capture each biospherian in his or her specific field of interest, and he collaborated with Barbara Smith, a performance artist from Los Angeles, on several interactive media events using video phones chronicling Barbara's journey around the planet and Roy's journey inside Biosphere 2.

With the exception of Linda (who didn't want one), we all have TV/VCRs in our apartments. Movies were the most common way to relax. Various members of Mission Control rent movies and then pipe them in to us. In addition, seven of us use the TV/cable system. The only source of frustration was our 'video-lag time.' We heard from friends about great new movies and had to wait until they finally released to video. Each apartment also had a boom box with AM/FM radio, tape deck, and CD player.

On one occasion we heard that a Star Trek episode aired featuring an episode about a colony of humans from Earth who were living in a biosphere situated in the polar regions of a barren planet. On a Saturday night, we watched a tape of the show on the big screen in the command room with rapt attention. The colonists, who had been genetically engineered to fit into their world of several thousand people, suffered from lack of communication with the outside world. They adhered rigidly to a program formulated two hundred years prior by their founders and were ignorant of current advances in sciences and technology. This regression caused a revolution. Some of their members wanted to re-join humanity by hitching a ride on the starship Enterprise. This portrayed a situation entirely unlike that in our biosphere, where one of our main objectives was to facilitate daily communication flows with the outside world. But the Star Trek episode made us contemplate the effect Biosphere 2 was already having on the human imagination.

Since our dinner companions were the same every night, we tried to vary the agenda and locations. Initially, we made Tuesday night dinners a cultural evening. For a change of scene, we ate in the Club, the mezzanine room that overlooks the dining room. For the first six months we enjoyed a series of discussion topics that rotated between the world of the great explorers (Columbus, Burton, Heyerdahl, Bates, etc.); art (performance art and current art shows in New York); and a series on mythology narrated by Joseph Campbell. Some nights were dedicated to topics relating to other scientists' work that complemented our research inside Biosphere 2.

On other nights we brought up microscopes and explored the intricacies of flowers brought in from the wilderness biomes; or we entered the world of the insect, examining our major adversaries such as the broad mite and thrip; and also our defenders: predatory ladybugs and their larvae which fed on the pests.

Some Tuesdays, we watched topical videos. They ranged from a "History of Insanity," to a documentary on the water and electric utilities of New York City. This last PBS special was fascinating as we compared planet Earth's present 'use and dump' approach to water and sewage with the closed-loop recycling of our new world. The enormous amount of sewage generated every day from the people of New York City could fill Yankee Stadium several times over. The sewage is hardly processed at all before it is piped out to the ocean. By contrast, we produced no trash at all; for there was no place to put it. Our sewage system was designed to be a carrier of valuable nutrients that we recycled back to our agricultural soil. Although Biosphere 2's small atmosphere, ocean, and quicker cycles necessitated ensuring total recycling, even with its larger reservoirs and slower cycles, adopting similar approaches to regenerating and recycling nutrients and water is becoming a necessity to keep Earth's biosphere healthy.

On Thursday nights, Mark Nelson hosted several study series over the two years. We did one on the psychology of small group dynamics, two on American literature, and several others. Roy initiated a series of autobiographical nights where we told our life stories, one each night. This fascinating weekly event continued through December 1991 and January 1992. Doing this autobiographical review helped to remind us where we came from and to articulate where we hoped to go.

Friday nights we relocate to an outdoor venue. We christened the north balcony that overlooks the IAB Café Visionaire and ate dinners at three cafe tables enjoying free-wheeling conversation while the sky transforms from a vivid red sunset to stony black night. While dining al fresco, from time to time we ordered from the attentive French waiter a cappuccino with an eclair or a brandy. Anything could be had at the Café Visionaire—with sufficient imagination.

Saturday nights had no set schedule. But Sunday nights we ate small portions of meat, toast the event with savory fruit juices, and gave speeches or presentations to bring our companions up to speed on any projects we had been working on. Not everyone takes part; some prefer their privacy. Few organized events interest all eight biospherians. A diverse crew with diverse tastes, some one or two nearly always elected not to attend a given event. But feasts and birthday parties were unanimously appreciated and always guaranteed a full house.

PARTIES AND FEASTS

The first Thanksgiving meal marked the beginning of a tradition of feasts which became an integral part of our culture during the two years. A feast is more than a good meal, it must include superb cuisine, and lots of it, so that the guests can engage in conversation for hours on end while eating to their heart's content. Birthdays gave us a reason to both feast and celebrate. Gaie had the first birthday inside, October 12, with the first Biosphere birthday cake, made with wheat flour, figs, and bananas, and an icing in the shape of an octopus made of frozen papaya and yogurt. Linda's birthday on November 11 featured a great ginger cake with banana yogurt and the first home-made pizza of the closure. Eighteen months later, in March 1993, Jane's and Laser's birthdays became mega-events that included three superb meals during the day highlighted by our precious rationed coffee and home-made ice cream. There was so much food that everyone was able to store some snacks for the following days.

During the second year, choosing a theme and a location for birthday dinners evolved into a favorite game. On Laser's second birthday, we had a campfire picnic on the beach, even though fires are prohibited in Biosphere 2. He hooked up a TV monitor on the beach where he played a video of a fire burning slowly until it became glowing embers. Some sat around the video screen enjoying our 'fire' while some took a swim in the coral lagoon. Jane's birthday was dedicated to those who work out with weights. We all assembled for the evening meal in the gym room dressed as muscle builders, then later took a sauna with hot steam generated from a little steam pressing machine above which blankets were hung like a teepee. Linda's birthday was centered in her bedroom to tell stories about our favorite dreams. Gaie's birthday was a cybernetic party in the command room where we mounted four video cameras from different angles to monitor the party and played a series of video channels oscillating between Michael Jackson, Madonna, and MTV displayed on the big six-by-eight-foot screen mounted above the table. Taber's July party was in costume; Mark's May birthday was an Australian Outback cowboy affair; Roy's, in June, was an Indian *lungi* party; and Sally's was a formal black-tie dinner.

For Gaie's second birthday inside, Roy had the clever idea of using alcohol from the medical supplies to spike the birthday punch! We all guzzled it down. Now was that desperation—or was it just following the doctor's orders? This lemon/orange ethyl alcohol punch had everyone looped and dancing on tables. It was one of our wildest parties, and we needed it. We had become acutely aware of the importance of good food and stimulants for boosting morale. In a continuing attempt to provide the essential inebriants, Sally made *chung* (a Tibetan rice beer) and banana wine, both of which were delicious. Our red beet whiskey was not. But aside from that, there was no alcohol to be had. Good food, plenty of food, and a variety of foods was absolutely essential for extended isolated missions, but let's not forget that stimulants, such as coffee, and inebriants, such as alcohol, loosen everyone up and promote conviviality.

Aside from birthdays, we partied on equinox and solstice days and, in general, on any holiday we could get away with! Three-day weekends are important morale-boosters, since we have far too much work to even think about anyone taking a whole week off. In months like August, which lack a biospherian birthday or public holiday, we declare a three-day weekend anyway because it's too long between July 4 and Labor Day!

Solstice and equinox became significant events because they marked the shift of seasons and a change in sunfall patterns. On these days we not only lived it up but contemplated the different routines we must follow with the new season. On our first solstice, December 1991, we celebrated a 'photon feast', glad that we had reached the shortest day of the year. The amount of sunfall would now increase for six months, making our wilderness ecosystems and our agriculture more and more productive. Languidly sprawling on the couches in the plaza, we toasted the sun every time it peaked out from under the clouds.

Sweet potato pies, cheesecakes, trifles, roast pork, a big pot of beans or posole, and coffee (if available) became the traditional party favorites. Here's the first Thanksgiving menu:

Baked chicken with water chestnut stuffing
Sautéed ginger beets
Baked sweet potatoes
Stuffed chili peppers sautéed with goat cheese
Tossed salad with a lemon yogurt dressing
Orange banana bread
Tibetan rice beer (*chung*)
Sweet potato pie topped with yogurt
Cheesecake
Coffee with steamed goat's milk

This menu was modest compared to future feasts which often included three or even more desserts, several types of bread, and sauces or chutneys.

Preparing the feasts was part of the fun, everyone got involved. Sally weighed and measured the whole foods used for each delicacy and generally baked all her treats the day before. Linda, who once made a Thanksgiving 'turkey' out of beans for a vegetarian group she lived with, became the number-one vegetable chef. Taber, famous for his patience in spending hours roasting pigs over open fires during pre-closure parties, took on the task of dressing the chickens or pigs, and the rest of us would dance around one another, chopping, mixing, grinding, and tasting in the crowded kitchen.

How we savored the thought that we would soon have more than enough to eat! No matter how well we adapted to natural foods, we still craved more sweets and stimulants as well as the massive amounts of protein and calories that a normal American diet includes. At the breakfast table, we'd fantasize about a spread of eggs, sausage, butter, hash browns, buttered toast, and coffee with cream and sugar instead of porridge, mint tea, and some mashed potatoes. It became a pastime to contemplate where all the food you can buy in supermarkets comes from and where all the trash associated with it ends up. In our little world, this type of opulence and waste was not possible. But on feasts days, feast we did!

To celebrate the anniversary of our first year inside Biosphere 2, Sally baked an enormous cake. Outside, another cake was readied. At 9:30 AM that Saturday, Margaret Augustine (Biosphere 2's CEO) cut the outside cake just as Sally cut into ours. Theirs was a delicate carrot cake, topped with whipped cream and decorated with a frosting tracery of the Biosphere. Ours was a Mack truck, a heavy, three-story job that Sally could barely manage to hold during the celebratory speech. Mark took the opportunity to weigh his piece—close to a pound and three-quarters. That meant the whole cake weighed fourteen pounds! The giant pieces of cake on our plates, we were told, made quite an impression against our slimmed-down frames. Make a real cake with heavy-duty ingredients like whole wheat flour, ripe bananas, and sweet potatoes, and then top it with thick slices of fig, papaya, and bananas and you have a dish fit for a hungry biospherian. Our friends kidded us about how fast those enormous slices of cake disappeared.

Some of the visitors on that day asked us if we missed quick, convenient food, such as hamburgers or pizza. Well, we did have pizza—it's often the special dish that people requested for their birthdays. Delivery is a bit of a problem: our pizza took four months to make. That's the time it took to grow a crop of wheat for the flour crust. Then we add the time for making cheese from our goat's milk and growing the tomatoes, green chilis, and eggplant—without even considering the time and effort necessary to produce any sort of meat topping.

On the Fourth of July, we gathered in the tower library to see fireworks from all sides, the distant ones in Tucson and the Biosphere 2 fireworks out on the grounds. And on New Year's Eve, we climbed the library stairs again to play poker until midnight. Our play was amateurish at best—in the end, Jane bluffed Taber into a large pot, but then lost because she'd gotten confused about whether a straight beats three of a kind or the other way around. None of us would be a threat in high-stakes poker in Las Vegas.

ARTS FESTIVAL AND ARTISTS

A high point of our initial year inside was the first Biospherian Arts Festival in April 1992, a feast for the mind, heart, and stomach. Sally made the first presentation—a wonderful afternoon tea, served on a table beautifully deco-rated with rainforest flowers. She served demitasse cups of mint tea, along with banana bread, goat's cheese, and banana/papaya jam. While we ate, we made presentations of our works in progress. These included video documentaries, photographs, paintings, poems, and other writings. Jane and Taber teamed up as an electronic band, complete with synthesizer, drums, voice, and pre-recorded 'sounds of the Biosphere'.

Inspired by the unique qualities of our world, Mark began to try his hand at poetry. Encouraged by Roy, who was also writing poetry, Mark hooked up with Roy and some poets at the Electronic Café (a coffee house in Los Angeles) for a joint poetry reading. The contrast between the two frames of reference was striking. Their poems were written in the aftermath of the L.A. riots and reflected the harsh, urgent tempos of the city. Mark's reflected a world where an undercurrent of pumps, sumps, and air handlers underlies a melodic chorus of crickets, the hoots of galagos, the rustling of wind-swept vegetation, and the sounds of a small band of humans who live in another world.

Our second arts festival was in August 1992 and this one was 'interbiospheric.' Jane decorated the interview room to look like a discotheque. We invited artists on the SBV staff (Marie Harding who exhibited her recent series of paintings of the Biosphere) and from Tucson (international touring classical guitarist Bill Matthews and artist Peggy Doogan from the University of Arizona) to participate. We set up a video connection to the conference room of the Biosphere 2 Inn and a linkup to the Electronic Café in Los Angeles. Musicians there jammed with Taber and Jane, our Bio Band, Roy and Mark read their poetry, and artists and guests in both worlds shared a wonderful evening.

In August of 1993, with under fifty days left for us in Biosphere 2, we held a second Interbiospheric Arts Festival, which was also a great success. We hoped future crews would continue the tradition.

STRANGE ENCOUNTERS

It must have been about six o'clock on the evening of our first Thanksgiving inside when over the radio came the oddest sounds of trumpets and horns. We dashed down to the big recreational space on the first floor where we saw a troop of people gathering with flashlights. The management staff of SBV had all gathered, dancing, singing, playing music, and wearing the oddest assortment of clothing. They looked like time travelers visiting from the future. We gazed out into the darkness with astonishment, our breath fogging the windows.

We wondered if we'd have Halloween guests come for a trick or treat and sure enough on the first Halloween, some friends did show up. We heard that the most popular Halloween costumes in Tucson that year were groups of eight going around dressed up as biospherians and that one group included a girl with a bandaged finger.

The full moon seemed to really get us going. Everyone seemed wide awake on these nights. It was common to get up for a walk around the Biosphere and meet someone wandering the hallways or paths, or even meet night-owl friends roaming outside of the Biosphere peering in!

One of the most interesting (and distant) links we made was in March 1992 with the twenty-two-person American research team at the South Pole. Part of our fascination was finding a mirror through which to see ourselves. Were we like them? What kind of explorers were we? The connection was achieved via phone from the Biosphere to a ham radio operator in South Dakota who was connected with members of the crew packed into their radio shack.

They had been there for over five months, and as we spoke, it was the eve of their entry into the six months of night when the sun dips below the horizon. They told us it was minus seventy degrees Fahrenheit outside with a wind chill factor of minus one hundred forty. We regaled them with tales of our lush tropical world in Arizona. Planes can't land for eight months out of twelve in Antarctica, so the videos and books on Biosphere 2 we sent them had to await their June mid-winter mail drop. We both lived in domes, but for obvious reasons theirs aren't made of glass. Evidently their main building, a fifty-foot-diameter geodesic dome, is on shifting and settling ice, so they were constantly chocking it up to stabilize it. They had a station cook who worked six days a week; they foraged for themselves on Sundays.

Their individual quarters were tiny—six-by-ten-foot rooms, which housed two crew members during the summer months. The doctor and station head have bigger quarters because their rooms were also used for medical facilities and emergencies. At the time we spoke, they were just setting up an eight-by-ten-foot space as a greenhouse, using hydroponics and high sodium lights to grow winter vegetables. They also did home-brewing like us, although they had a good supply of beer. We survived quite well without beer, but we weren't sure we could handle six months without the sun!

FAMILIES AND FRIENDS

You can prepare for saying good-bye, but in some ways no preparation is enough. Some biospherians wondered what would happen when they physically separated from their partners for two years. And what about families? It took some getting used to, but we made our family lives thrive by means of window meetings, telephone, email, electronic faxes, two-way radio, and video technology. PictureTel, an interactive video system we had inside, relayed real-time images and sound from any two parties that had the same system, just like a satellite linkup, but it's done through special phone lines. The first Christmas in the Biosphere, all the biospherians were linked up to their families for a one-hour meeting, whether they were in London, Antwerp, California, or New York.

Mark made a Chanukah linkup with his mother, Minnie, in New York who had his aunt and two friends there to chat with her son. These were women in their seventies and eighties who had come to America from the Old World and were visiting by way of a futuristic high-tech medium. Mark enthralled them with a nine-minute video Laser made of life and work inside Biosphere 2. What struck them especially was the contrast between the labor of growing our own crops and our high-tech kitchen. They enthusiastically suggested some traditional Jewish recipes to add to our repertoire. Minnie sent him a recipe for a sweet potato pie (kugel) and his aunt sent one for gefilte fish which we could make using our tilapia fish. In the long run, our contacts with our families and friends became the magic ingredient that made Biosphere 2 a place to call home.

*Mark Van Thillo (Laser) contemplates re-entry into Earth's biosphere
from the library tower.*

The Adventure Never Ends

"I believe in the research program initiated here at Space Biospheres Ventures . . .
I see Biosphere 2 as a shining beacon pointing the way to an expanding future for
humanity . . . it is contributing to the critical next step: closed ecological systems
that will expand terrestrial life throughout the Solar System."

– **Dr. Thomas Paine,** Chairman, National Commission on Space,
Administrator of NASA during the Apollo Project, 1990

CHANGES IN A NEW WORLD

AFTER EIGHTEEN MONTHS we all agreed that we really felt at home, happy in our
adaptation to this new world. It wasn't easy, but we had finally adjusted to a new
diet, a new work schedule, a different social milieu, a dramatically different atmosphere with falling oxygen and fluctuating carbon dioxide, and the unique seasonal
patterns of our biomes. And just then, it was time to prepare for our departure and
to hand over our activities to a new crew.

At the second New Year's Day we celebrated in Biosphere 2, Laser commented
that sometimes his body seemed to adapt and other times it seemed to fall out of
synch. Dr. Oleg Gazenko, a pioneer in Russian space medicine, succinctly described
the acid test of successful adaptation to new conditions when he visited Biosphere
2: "You will know your body has adapted when you feel a sense of freedom." By
March of 1993 many of us independently experienced a distinct change. There
was a deepened feeling of unity among the eight, a feeling of rest and well-being.
Eighteen months had been a long haul, with many ups and downs, but with the
realization that we would now probably complete the mission, it indeed was a
feeling of freedom.

In checking the literature on small groups in isolated environments, Roy found
that frequently the third quarter was the most difficult. The fourth and final quarter
was usually far easier, with high morale returning in anticipation of completion.
This matched our experience in Biosphere 2. Our third quarter had been the one
with the severest oxygen depletion, the dark and stormy winter with its impact
on both food crops and our good spirits, and the greatest disharmony among the
crew. By contrast, our last six months were marked by sunny skies, high morale, a

deepened sense of physical well-being, and an emotional harmony that could ride the ups and downs.

Future biospherians may find it easier, since we spent much of the first year debugging the system, finding out what would go wrong or be the greatest challenges. Our second year shifted to fine-tuning systems, operations, and expanding research endeavors. And, of course, we also had more time to relax and absorb the experience. It was a time for discovery and contemplation of our role in this new world. We gained the detachment to appreciate the enormous effort which the two years had demanded. It seemed as though we had touched every aspect of our world; we interacted with molecules and with trees, we knew our environment's boundaries and its subtleties.

We anticipated our departure with nostalgic sadness. There was no doubt in our minds that at 8:15 in the morning on September 26, 1993 we'd struggle with the departure. It would be a sharp transition, even though only an airlock door separated the two worlds. It was as if we had departed the planet for two years, and in some sense, we surely had.

HANDOVERS AND UPGRADES

It was in our last six months that we were forced to get deeply involved with the planning for Transition One. This is the transition phase between this crew leaving and the next mission's crew starting. Gaie spent many hours a day working with John Allen, Vice President of Biospheric Development, and Jack Corliss, the Director of Research, on the organization of the transition research and development. Sally worked with Norberto Alvarez-Romo, Vice President of Mission Control, to develop the biospherian training program and selection process for future teams, while organizing the process by which the crew would hand over operations to the new biospherians. Laser and Larry Pomatto, SBV's Director of Technical Operations from Systems Integrated, organized the technical changes that would be required during the transition.

The transition would be a challenging time because, among many factors, we aimed to maintain the integrity of Biosphere 2's atmosphere. That meant that people were permitted to enter the Biosphere using the airlock system so long as they do not exceed 192 population-hours per day. This number of hours was equivalent to having eight biospherians live in the biosphere for twenty-four hours. The amount of carbon dioxide released, and oxygen consumed, would be the same. This procedure ensured that our long-term monitoring of atmospheric cycles could continue practically undisturbed. Because there would be considerable changes made in technical systems, the transition was expected to take five months. We anticipated that the focus of transition's final two months would be on the next crew and their final training before they began their mission. We thought that future transitions between missions could be reduced to about six weeks for research and operation handovers.

In taking Biosphere 2 from theory to practice, we gained clear perspectives of what worked and what didn't. The few technical systems that would be replaced almost all involved simplifying the original system. Computer-controlled systems have a tendency to become more complex than necessary. In Biosphere 2, we learned to simplify the design, so it accomplished only what was necessary.

Sally was in charge of teaching the new team the operating techniques for the entire system. Although the Mission Two team had not at that point been finally selected, the pool of applicants started their training on site during our second year of closure. It was an extraordinarily culturally diverse set of men and women. Candidates again ranged from twenty-year-olds to those in their sixties; hailing from many nations, including the former Yugoslavia, Australia, Mexico, Nepal, Germany, Britain, and the United States. Choosing the right mix of people was vital because they must possess skills diverse enough to handle the hundreds of systems, the knowledge to carry out a wide variety of research projects, and the ability to interact with the range of people interested in Biosphere 2, from scientists, environmentalists, to elementary school children.

During the tight timetable for people allowed into the Biosphere during transition, research scientists had high priority to make complete surveys of insects, animals, plants, soils, waters, and introduction of new species. Every plant in the wilderness biomes would again be tagged and measured, which enabled us to record growth and losses. In the ocean, we would not only make some technical upgrades but would also enhance the food web to include more herbivores (fish, sea urchins, and snails) to handle the proliferation of plants (micro- and macro-algae). All the colonies of corals being tracked would be tagged and mapped to follow them through time. We were to add roughly twenty new woody trees each to the desert, savannah, and marsh.

In the rainforest, a major transition objective was to increase the number and variety of edible species. In particular, adding more productive types of bananas, taros, papayas, and other distinctive rainforest food plants. In the cloud forest we would make attempts to match a new plant mix to the environmental conditions there. With the planned removal of our initial tree canopy species, we could also increase the biodiversity of the forest floor and lower canopies of the lowland rainforest. In the agriculture biome, we would enrich the diversity of low-light and understory crops and include more supplemental artificial lighting in shady areas.

We biospherians were no longer permanent residents of the Biosphere during transition, but we remained quite involved in it as we entered and exited through the airlock to transfer our knowledge to others. When we moved out on September 26th, our live-in crew of eight was replaced by a single, rotating, nighttime watch member during the transition phase until the crew of Mission Two moved in. During the day, Mission Two crew did the agriculture and other basic operations for about four hours per day.

THE IMPOSSIBLE HANDOVER

We could hand over the controls, but it would be almost impossible to hand over our experience. We took for granted the cycles of water and the composition of the air we breathe—until we faced our daily role in their maintenance. This was a totally new experience, a way of life we discovered by living inside an artificial biosphere. Out in Biosphere 1, every action has its consequences, even if we have difficulty understanding them.

It was the founder of biospheric theory, a Russian geochemist named Vladimir Vernadsky, who was among the first to see, in the early 1900s, that humanity was altering Earth's biosphere with the expansion of cities, industry, technology, travel, agricultural production, and pollution. "Man," he said, "is a powerful geological force." Vernadsky had the vision to observe, long before the environmental crises of our time, that we needed to evolve in order to reconcile the conflict between technology and ecology. Humans must work together and become intelligent stewards of the biosphere or face the threat of irreparably harming its ability to sustain life. "This," said Vernadsky, "will be the beginning of our Noosphere."

Man's invention of tools and technologies is an extension of life's inherent drive to continually extend its domain. Our invention of spaceships, coupled with our ability to create biospheres, may enable life on Earth to expand into the solar system and beyond in the millennia to come. This is exactly what Vernadsky suggested in his far-reaching conclusions. Biospheres are a cosmic phenomenon, linked to the universe through our dependence on energy from the sun. But far from being confined to any originating planet, biospheres in the fullness of time will spread life throughout space. Biosphere 2 was undergoing an evolutionary process, one partially molded by those who lived inside it, but the people are a part of that evolution as well. Our collective challenge is to learn to evolve in harmony with the biosphere instead of in conflict with it. This is not easy—it will require profound shifts in our thinking and in our actions.

LEAVING

Although we chatted from time to time about leaving, none of us really grasped its impact. How could we anticipate what the shock of returning to the outside world would be like? Physiologically and psychologically there were major adjustments, just as there had been when we entered. We hadn't known what to expect then, either.

In some ways, we have been renewed by our experience. Our cholesterol levels were at extraordinarily low levels. Our blood pressure lowered and our pulses slowed. Complexions were clear. We had minimum exposure to ultraviolet radiation, were eating remarkably fresh and uncontaminated foods, breathing biologically cleansed air, and drinking pure water. Yet mastering life inside Biosphere 2 had been a struggle, and every step of the way was hard-won. So who are we, now that we approach re-entry, and do we even fully realize the changes we have undergone?

Mark made a speech during his second birthday celebrated inside Biosphere 2. He concluded with an acknowledgment that he was, quite simply, happy. It was amusing to hear our erstwhile number one grouch declaring life to be happy and good. Even after all the sweat and tears, and not a few heated discussions, we had become a united team dedicated to the task of keeping our little world in fine shape.

FOOD FANTASIES AND FAMILY SUGGESTIONS

Part of the preparation to leave involved deciding where we were going to live. Some, feeling the desire to travel again in the world, put some of their energy into getting cars so they'd have wheels the first day out! There was a great deal of discussion about buying electric cars to help reduce fuel emissions, but in reality, there was no such car on the market capable of covering long distances. Sally shopped for a black evening gown to wear during the first dinner which was to be held on the green lawn just outside the habitat on the night of re-entry. Her friend brought the latest Spiegel catalogue to the window so she could order her dress in advance!

All of us dreamed about which foods we would devour first. Taber announced one morning at breakfast that he was going to ask his father to bring a giant chest of food so that during the first day anyone could discreetly disappear for a minute, delve into the chest, and grab some goodies. Carole Hemingway, our media consultant from Los Angeles, promised to bring Laser a Snickers bar cheesecake (he heard it was a California specialty). Gaie said she wanted champagne—a case would do—and Sally's mouth watered for lox and bagels. Mark dreamt of a breakfast of freshly baked croissants and a cappuccino. Roy, who was forever trying to entice someone to take pity on the biospherians and convince Mission Control to allow in a case of Scotch, wanted a bottle of his favorite brand waiting for him when he stepped beyond the airlock door. In a departure from the food fixations of the rest of us, Linda most looked forward to being with the people dearest to her and a trip into the wild, wide open spaces.

Many of us felt a concern, or perhaps just curiosity, about what it would be like to interact with all the people outside! We had the luxury of great privacy inside the spaciousness of Biosphere 2, where it was rare to pass others in the hallway, and we carried two-way radios to stay in touch. But on the outside, we'd constantly be around people who say hello and want a response. There were major adjustments to make in the fast-paced, people-filled world of Biosphere 1. We looked forward to ending our isolation, but at the same time we realized, with a bit of trepidation, that we no longer had the walls to protect us.

PERSONAL PREPARATION

We anticipated that Biosphere 2 would be as much a time machine as a journey in finite space. The ceaseless transformations and movements of matter were quickened since life and technology were so concentrated. Time seemed to have expanded at the same time as our space had condensed. For example, since our air and water cycles were hundreds, if not thousands, of times faster than the equivalent Earth cycles, one Biosphere 2 year was a far greater time-event than an Earth year. Perhaps this is one reason why our experience in Biosphere 2 was so rich and deep.

Rusty Schweickart, a crew member on Apollo 9, served on the original Project Review Committee of Biosphere 2. Rusty eloquently described his experience watching Earth from space as an orbiting globe, a unified whole without the national divisions that politics and mapmakers impose. At the first Biosphere 2 workshop in December 1984, he shared an enlightening perspective about his space experience:

> "You are up there as the sensing element, that point out on the end, and that's a humbling feeling. It's a feeling that says you have a responsibility. It's not for yourself. . . And when you come back there's a difference in the world now. There's a difference in the relationship between you and that planet and you and all those other forms of life on that planet. . . It's a difference and it's so precious. And all through this I've used the word 'you' because it's not me, it's not Dave Scott, it's not Dick Gordon, Pete Conrad, John Glenn—it's you, it's we. It's Life that's had that experience."

When someone asked Rusty how he had prepared to go to space, he answered that he prepared for his momentous task by reviewing the great thinkers of the world, the poets and philosophers who spoke most clearly to his heart. Joe Allen, the shuttle astronaut, whose Midwestern, Will-Rogers sense of humor and drawl disguised the fact that he holds a PhD in physics from Yale, had quite a different perspective. When asked the same question, he replied, "The night before my flight, I studied the operating manuals."

Like Rusty, we felt that we were granted the privilege to experience a new relationship to our global biosphere by inhabiting a man-made one. Like Joe, we know that without operating manuals we cannot do our job. Perhaps these are the most important approaches that we can use to share our experience with others.

In his analysis of the core mythology that lies behind all human culture, Joseph Campbell traced what he calls "the hero's path." The first stage is the vision or calling. Then comes the adventure in a new world: the land of magic and mystery, where many obstacles must be overcome but great wisdom can be won. The last, and perhaps most difficult stage, is the return. The hero must re-enter the ordinary world and help change it in accordance with the insight gained on the vision quest. While we do not view ourselves as heroes, this powerful sequence has great resonance for us. Our greatest challenge is yet to come—to communicate the reality we experienced living under glass.

At the end of its two-year maiden voyage, Biosphere 2 is like a well-tested ship, ready to carry on research in biospherics for decades to come.

AFTERWORD

RE-ENTRY

"Biosphere 2 will really lead to a tremendous increase in our understanding of environmental problems and that in turn will lead to new technologies to heal the world and make it a better place for our children and their children tomorrow."

– **Jane Goodall**, primatologist and anthropologist,
Re-entry day speech, September 26, 1993

ON THE EARLY MORNING OF SEPTEMBER 26, 1993, we packed our bags and moved out of our Biosphere 2 apartments. With very mixed emotions we waited to leave near the inner airlock door where we could hear the ceremony commencing. Gathered outside were about 1,500 people and over 10 satellite-dish vehicles broadcasting the 'return of the biospherians.' The Tucson mayor proclaimed that day "Biosphere 2 Day" and a Tucson orchestra performed Vivaldi's "The Four Seasons." Then, it was time to depart our home after 731 days.

An array of distinguished scientists who had visited and worked with us spoke. Among them: Jane Goodall, the chimpanzee expert; Sylvia Earle, renowned deep ocean explorer, Chief Scientist for the National Oceanographic and Atmospheric Administration (NOAA); Oleg Gazenko, the head of the Institute of Biomedical Problems, Moscow; and Harold Morowitz, a distinguished biophysicist from George Mason University. Additionally, John Allen, Margaret Augustine and Edward Bass, the three people who were most instrumental in creating Biosphere 2. Lastly, the eight biospherians spoke about the significance and personal meaning of the two-year experiment.

As soon as we stepped out of the airlock door, we were astonished to experience how different it felt being back in Biosphere 1, our planetary biosphere. It surpassed what we imagined when we had contemplated this moment with mounting excitement as the months and days moved us towards departure. Those first breaths in our new world: a different atmosphere, thinner and lacking in the rich tropical and organic smells of our mini-world. The wonder of seeing the horizon and sky, with its subtle palette of colors, without glass and a space frame overhead. Feeling the physical and emotional closeness with those

213

who came to celebrate our emergence was overwhelming. No longer were we communicating through a glass barrier or over telephone. We had fun at the obligatory press conference packed with journalists from around the world; it was startling to learn that close to a billion people watched the re-entry via the new satellite TV coverage. Later, the doctors periodically called us in for physical exams and to see how we handled being in the midst of so many people and so much food! They measured our body fat content and found we were all comparable to professional athletes.

We climbed into a golf cart after the ceremony was over to go to Mission Control for a lavish lunch buffet. It was strange to eat food which we hadn't grown, even if some had fueled our periodic Biosphere 2 food fantasies. The gasoline engine powered golf cart was equally bizarre: we immediately noticed the horrible exhaust smell and how strange it was to be back in a vehicle. The evening dinner featured a live satellite link-up from Sri Lanka with Sir Arthur Clarke.

We laughed as media coverage of our successful completion of the two-year experiment was highly congratulatory, even from outlets which had previously been quite the opposite. Time Magazine declared the project one of the ten best science projects of 1993.

The next days and weeks became a blur as we said goodbye to life as we knew it with only us eight and our biosphere. Indelibly etched in our minds was our first trip to a supermarket with endless food aisles packed with cans and packages and lists of exotic chemicals on their labels. Driving a car to town to get food with cash or a credit card without having to plant, tend, harvest, and process the crops it came from. Not to mention all the packaging, cardboard and plastic which accompanies store-bought food in Biosphere 1. This old reality was confusing since we'd been absent for so long. Re-engaging with our old lives, we became aware of how removed modern life is from ecology, from nature. We had lived in Biosphere 2 with daily concerns such as "is our water drinkable?" and "where are our wastes going?" and knowing what was in our atmosphere and where our food came from. Now, this was no longer part of our lives. It took time to fully adjust back to our Earth's biosphere while we still vividly recalled how we had lived inside Biosphere 2.

TRANSITION

For many of the biospherians, our first period outside Biosphere 2 involved helping with technical upgrades and detailed research projects in the period between our departure on September 26, 1993 and the entry of a second crew of biospherians on March 6, 1994.

Gaie, as Biosphere 2's Director of Research inside, was in charge of the transition. That period of time included coordinating research, technical improvements to the facility, training the second crew, and upgrades to computers, controls, and archival systems. The airlock doors were closed aside from people entering and leaving. Our aim was to keep Biosphere 2 operating as a closed system (while simulating the number of people inside the facility to approximate a crew of eight), and thus ensuring the integrity of the atmosphere and minimizing impacts on the system.

Transition research included a resurvey of the coral reef ocean to map the living corals, fish, and invertebrates; aquatic and marine microbiological surveys; a complete resurvey of the wilderness plants; review of the agriculture with a close look at integrated pest management and new cultivar selections; entomological surveys; genetic sampling of two species of *Gambusia* of fish in a marsh experiment who appeared to have become one population; sediment cores; oxygen and carbon isotope studies; and much more.

Mark managed the overall mapping and measuring of the terrestrial biomes as he had done pre-closure working with faculty and students from the Yale School of Forestry and Environmental Studies. Linda worked on the tropical rainforest survey. A total of over 11,000 plants were detailed, including ones which died, as well as new plants which started during the two years. It's a remarkable dataset documenting the changes and ecological self-organization that occurred since the plants were examined prior to closure. Sally worked with Dr. Richard Harwood of Michigan State University, Dr. Jim Litsinger and other consultants to improve agricultural production with new crops better suited to Biosphere 2 conditions, as well as an enhanced integrated pest management program. Improvements included the addition of supplemental lighting for farm areas with reduced ambient light.

In resurveying the ocean, Dr. Phil Dustan and Dr. Judy Lang worked with Gaie to map a total of 917 hard and soft coral colonies (as compared to the 1,266 which we began with pre-closure) that were still alive and adapting to their new world. Only 25% of the corals died during the two years and there were 38 new coral recruits along the shallow walls of the wave machine and tank, as well as on the reef. Dr. Donald Spoon, from Georgetown University, led a resurvey of the streams and marine systems in which he tallied over a thousand microbiota. He concluded that these ecosystems closely resembled natural systems and that the high biodiversity of these interconnected biomes boosted the system's ability to respond to the environmental changes of the two-year closure.

Laser, Manager of Technical Systems, oversaw all the upgrades of the technical systems that had caused problems or required excessive maintenance during the first closure experiment. The tasks ranged from repairs to the replacement of entire systems, such as the installation of a new water pressure tank, new planter boxes and lights in the agriculture, 20 new ocean skimmers to replace the few that had been built inside during the two years and electronic/cyber systems improvements as well as an overhaul of all the control systems.

Sally was in charge of the training of the second team who were even more diverse than the first crew. It included people from Mexico, Nepal, Australia, Germany, the United States, and the United Kingdom. Her responsibilities included the management of the incoming crew and supervising the veterans of the first mission who trained their successors in operating Biosphere 2. We knew that truly becoming a biospherian steward would require the new team living on their own and experiencing how their actions were important to the health of their world.

We were proud and fulfilled to have left Biosphere 2 as a healthy, vibrant and beautiful biosphere; it was most rewarding to gift it to the next crew to experience. Now, we faced a new challenge: How can we incorporate what we learned in Biosphere 2 and lead even more meaningful lives as Earth 1 biospherians?

RETURN TO EARTH: OUR LIVES OUTSIDE THE GLASS

ABIGAIL ALLING (GAIE)

Because we had experienced that our coral reef was the indicator biome for the health of Biosphere 2, I was keen to apply this understanding to our Earth's oceans. Thus, with John Allen and Laser, we launched a 14-year voyage (1995-2008) around the world to map and monitor coral reefs with the Planetary Coral Reef Foundation. Over the course of this expedition, our data showed that two-thirds of the 48 reefs studied were at risk, and all showed signs of decline. Realizing that the coral reefs were an important index for ocean health, and thus our biosphere, we also developed a satellite to monitor and map earth's coral reefs in real-time. The Coral Reef Satellite Mission design was led by Dr. Phil Dustan and Robert Goeke (MIT), along with several other engineers and scientists, Scripps Institution of Oceanography, Astrium GmbH, and others.

In 1995, The Planetary Coral Reef Foundation also invited Mark Nelson to assist with the development of its Wastewater Gardens®—a subsurface wetland recycling system for grey and black water that uses no chemicals or pumps, provides food and flowers, and treats the wastewater ecologically while protecting coastlines, and other places that they are installed from pollution. The prototype for the system was the constructed wetland installed inside Biosphere 2. It was clear to us all that this technology was needed everywhere.

In 1999, The Planetary Coral Reef Foundation changed its name to Biosphere Foundation in order to embrace other projects, as well as develop practical skills regarding biosphere stewardship. Sally had already joined us in 1995 as our Chief Financial Officer and together with Global Ecotechnics Corporation, Biosphere Foundation initiated the design and development of a space biosphere for a manned mission to Mars. The project was called Mars On Earth® and began with building and then experimenting with light, soils, crops and atmosphere dynamics in a 12 foot diameter, opaque, closed system facility titled: 'Laboratory Biosphere.' That research provided us with critical information needed to create a sustainable, soil-based agriculture/horticulture system that could support humans off this planet.

The years from 1995-2005 coalesced the knowledge of the core Biosphere 2 team to develop these projects, both at Biosphere Foundation, as well as at Global

Ecotechnics Corporation and its numerous Institute of Ecotechnics projects around the world. We worked together to harvest our work at Biosphere 2 and continue to practice what we learned.

The economic and political climate of 2006 was not receptive to financing the development of a space biosphere, nor a coral reef remote-viewing space platform, so we turned our attention to local action and education. Today, I, along with Laser and Sally, lead Biosphere Foundation and its programs from our base in northwest Bali, Indonesia. There, we engage with community-led restoration programs for coral reefs and forests, assist farmers in testing new cultivars for a dry monsoon region, and install Wastewater Gardens® to re-use water for irrigation and recycle waste to reduce pollution. Daily, we apply what we learned about re-creating and stewarding Biosphere 2's biodiverse wilderness biomes with the aim to restore ecosystems and conserve biodiversity to enhance resilience to climate change. For example, our community-led "Coral Steward" training program teaches people how to use simple techniques, such as replanting broken, but living, coral fragments back onto the reef using concrete, just like we did all those years ago in the Biosphere 2 ocean.

As I write this onboard our 113 foot sailing ship named *Mir* (after the Russian space station that orbited the Earth for 15 years with cosmonauts who cared for its numerous species of plants and animals), Laser is back on a construction site, leading the completion of our new Biosphere Center in North West Bali and Sally is organizing our Biosphere Stewardship Education Programs to train new "biospherians" to become tomorrow's leaders. The Biosphere Center, in conjunction with our flagship research vessel *Mir*, provides a comprehensive two-pronged approach for stewardship, research, and collaboration.

As for me personally, I left Biosphere 2 in 1994 knowing that I had been given a unique chance to be part of creating and living in another world, and so discovered a new relationship with my biosphere. Because we had the ability to monitor the experiment in detail, and thus clearly see the results of our actions, we learned how to communicate with our biosphere. This became a co-creation between the crew of eight and our living system; a process that formed a feedback loop, which led us to become its constructive stewards. Additionally, the overall experience provided us with a microscopic view about human dynamics, which provides invaluable insights into the challenges we face today.

I know that, as part of the crew of our Earth's biosphere of around 8 billion people, that we share an uncertain future, just like we did during those Biosphere 2 years. Perhaps the first step towards solving the challenges ahead is to understand that we are part of a biosphere experiment happening right now on Earth, and we are each responsible for our actions and its outcome. I hold Biosphere 2 bright inside me, guiding my way. If it worked in Biosphere 2, then it can work for Biosphere 1. It is an on-going, daily conversation as I compare both biospheres and the experience of living in two worlds. We learned together, then, about how to be beneficial co-inhabitants with our Biosphere 2, and so can we all learn today. All we need to do is recognize if we are enhancing, or rapidly degrading, our Earth's well-being and then adjust our actions accordingly. We have the right to a healthy biosphere with plentiful clean water, a breathable atmosphere, nourishing food, and biodiverse wilderness systems. We all share our planet's biosphere and simply put, it's ours to care for and to love.

LINDA LEIGH

As the Director of Terrestrial Ecological Systems for the Biosphere 2 rainforest, savannah, and desert, Linda had led the work with our consultants to choose soils, plants and other species, including the galagos and insects, and build its necessary food webs. With a background as a field and conservation biologist, Linda had delighted in her role as a botanical investigator in the newest world.

Afterwards, she attended the University of Florida to pursue a PhD working with Prof. H.T. Odum, one of our most knowledgeable consultants, known as the "father" of ecological engineering and a founder, with his brother Eugene Odum, of systems ecology. Linda's dissertation topic was an examination of the fundamentals of biodiversity and compared Biosphere 2's rainforest with natural counterparts.

After completing her PhD, Linda returned to Arizona and taught ecology at Central Arizona College. There she sought to give her students a feel of closed ecological systems and developed what she calls "bio-tubes." These plastic-wrapped habitats enabled her students to live in a micro-world filled with plants and soils to get a personal experience of the interconnectedness of all life.

Now living in Oracle, Arizona, just five miles from Biosphere 2, Linda is an environmental consultant and, with her partner, started an organic farm and worm-farm enterprise called 'Vermillion Wormery'. Aiming at zero waste, Linda is developing vermi-composting systems using the power of earthworms to transform (what is ordinarily) organic garbage into soil-enriching amendments. She also occasionally gives talks at Biosphere 2 for visitors, students, and staff. One of her speeches is entitled "the Green Lessons of Biosphere 2."

TABER MACCALLUM AND JANE POYNTER

Jane and Taber were married about a year after leaving Biosphere 2 and the ceremony was held in the facility. Together, they went to the University of Arizona to complete their university studies and then quickly moved on to space exploration.

While living inside Biosphere 2, they had also started the Paragon Space Development Corporation. They enlisted aerospace engineers and scientists that Taber met while he completed his studies at the International Space University program, prior to his two years inside Biosphere 2.

Under their leadership, Paragon has developed technologies for life-support systems and thermal control to deal with extreme environments, like under the ocean or in outer space. Amongst their notable accomplishments was research with the first lab-sized closed ecosphere, which flew both on Mir Space Station and a Space Shuttle flight of NASA. This tiny world included aquatic plants, microbes, algae, and small crustaceans, which constituted the first closed ecological system ever flown and studied in the micro-gravity conditions of space. Aquatic animals were able to reproduce in space conditions—another significant milestone.

Paragon also set a world record with a safe descent of Alan Eustace, a vice president at Google, from a helium balloon at a height of 135,000 feet, reaching a speed of 820 miles per hour. This stratospheric expertise birthed a spin-off enterprise, World View Enterprises, offering opportunities for people to experience "near-space" and for commercial use of the high platform for communications and environmental monitoring.

Jane is currently an environmental consultant focusing on green technologies and sustainable development. She has worked on World Bank projects to grow alternative crops as a response to global climate change in drought-stricken areas of Central America and Africa.

Highlights of Jane's educational outreach include hosting a children's program filmed while she was on the parabolic flight of the so-called "vomit comet" airplane which gives a brief experience of zero-gravity conditions. She also hosts another TV program called "Going Green" which looks at ways of improving human health and the environment.

Taber is currently Chief Technology Officer of World View, the space balloon company. While at Paragon, he served for twenty years as CEO and Chief Technology Officer, supervising its technology developments that flew on the International Space Station, NASA, and Russian commercial space flights. He was core team leader and safety officer for the record-breaking, near-earth flight descent from balloon.

Taber, who not only ran the analytic laboratory inside Biosphere 2 but was also closely involved in developing the innovations to allow it to function without use of toxic chemicals, has continued with his technological wizardry. He holds patents in spacecraft heat control, diving in contaminated waters, spacecraft testing, and analytical chemistry, including a co-patent for some of the air monitoring developed by Space Biospheres Ventures for Biosphere 2. Taber was honored by *Popular Science* magazine which named him the 2008 Inventor of the Year.

Still very committed to space exploration, Taber conceived and helped design a NASA-funded, portable life-support technology integrated into a Mars spacesuit. He was Chief Technology Officer for Inspiration Mars which had planned to do a Mars "fly-by and return" mission with two people. Jane and Taber had been prime candidates to do that Mars mission, a journey that would have been comparable to their two years inside Biosphere 2!

MARK NELSON

One of the marvels of living in Biosphere 2 was the connection we had with nature and how we could engage with its ecological cycles. Because our world was so small, it was very clear that everything we humans do has consequences. I wanted to make systems to enable people to reawaken their connection with the natural world which keeps them alive and healthy.

I fell in love with the constructed wetlands that I managed in Biosphere 2 to treat and recycle human and domestic animal wastes. They provide an ecological solution to a waste product that produces environmental pollution and contamination of drinking water. What excited me most about wastewater treatment wetlands is the way they reconnect people to the fundamentals of their lives: where their

water comes from and where their wastes go. I returned to academia to study ecological engineering so I could implement constructed wetlands around the world.
After a M.S. degree in Watershed Management from the University of Arizona, I
studied with the master of ecological engineering, Prof. Howard T. Odum, at the
Center for Wetlands at the University of Florida.

I worked with the Planetary Coral Reef Foundation to build systems in
Akumal, Mexico. I decided to make these systems even more biodiverse and beauti-
ful than the ones we had in Biosphere 2. They were renamed "Wastewater Gardens"
(WWG). We built over thirty WWGs along the coast, protecting the corals where
Biosphere 2's were collected. The WWG systems turn a polluting waste into stun-
ning, lush, and odorless gardens.

After I completed my PhD, I continued working as head of PCRF's Wastewater
Garden Division, then in 2004 started a new company, Wastewater Gardens International, with regional affiliates around the world. We've now installed WWGs
in a dozen countries, completing over 150 projects in different climates and environments. Currently, I'm part of the "Eden in Iraq" project team designing an art/
ecology WWG for Marsh Arab towns in southern Iraq, working with artist Meridel
Rubenstein and Nature Iraq.

Excited by the potential of closed ecological systems for Earth and space applications, I've organized many workshops at COSPAR (International Committee
on Space Research) General Assemblies. I also serve as an associate editor handling
research papers in the field, first for Advances in Space Research and now, for Life
Sciences in Space Research.

After Biosphere 2, the core design team (with collaboration from Institute of
Ecotechnics, IE, and the Biosphere Foundation) designed and built a far smaller,
closed ecological system facility, the Laboratory Biosphere in 2002 at Synergia
Ranch in Santa Fe, New Mexico. This is where I first began my ecological work in
1969 (a 22-year old New York City kid) and helped found the Institute.

I have served as IE's Chairman since 1982, organizing our interdisciplinary
international conferences and consulting to a series of affiliated biomic projects.
These include a sustainable timber tropical rainforest project in Puerto Rico, a
savannah regeneration project in West Australia, the *RV Heraclitus*, our ocean-
going ship, and the October Gallery in London.

At Synergia Ranch, the ecotechnic goal is to create an oasis in arid high juniper/

piñon grasslands. In the 1970s, I ran an extensive tree planting and soil making program to reverse desertification. We planted a thousand trees including a four-acre fruit orchard. Now, I help our new younger management team operate Synergia Ranch Organic, producing fruits and vegetables. I re-experience the joys of Biosphere 2 farming, growing beautiful certified organic produce. Synergia Ranch was recognized for decades of land care with the Good Earth Award at the New Mexico Organic Farming Conference.

My experience in Biosphere 2 continues to inspire me. I publish popular articles and scientific papers on the ecological challenges of our times. My two books are *The Wastewater Gardener: Preserving the Planet One Flush at a Time* and more recently, *Pushing Our Limits: Insights from Biosphere 2*, where I collected its scientific results and lessons and reflected on the human dimension of living in a mini-world.

In 1994, I was honored to give a talk at the Royal Geographical Society in London, sharing our explorations in a new world. More recently, I gave talks at the Eden Project in Cornwall, UK and then in London with Sir Ghillean Prance at the Royal Botanical Gardens at Kew and the Linnean Society. We explored the history and deepening relevance of what Dr. Prance calls "one of the greatest experiments of the 20th century."

My two years in Biosphere 2 were the most profound of my life. The deep visceral experience we biospherians had realizing our connection to, dependence on and responsibility for our living world changed me. That amazing journey motivates me – and all of us - to share our experiences to inspire our fellow Earth biospherians to more fully comprehend their connection to nature so we will take much better care of our world. There is no airlock we can leave through. We need to rekindle our love of our planetary biosphere, then act to protect and regenerate it; after all, it is our life support system.

SALLY SILVERSTONE

I came into the Biosphere 2 project via the rainforest. In 1996, after what was frankly a mourning period for the loss of Biosphere 2 and the destruction of its unique research potential as a closed system, I returned to the rainforest.

Tropic Ventures at Las Casas de la Selva is an Institute of Ecotechnics project, based in the mountain highlands of Puerto Rico. The aim is to research enrichment of existing secondary forest with valuable tropical hardwoods, without destroying the ecosystem. Back in the early 80's I participated in the setting up of the project and the planting of some forty thousand trees. The project was a collection site for the rainforest of Biosphere 2 and I assisted with decisions regarding which plants we would collect. In 1985, I left Puerto Rico, and returned to Arizona for the second major Biosphere design conference, with a list of potential plant species in hand. I ended up not returning for 13 years! By the mid-nineties it was time to return to Puerto Rico and see how those trees were doing.

What followed was 10 years as Director of Tropic Ventures at its education and research foundation. The early years were quite a struggle but by the time I left in 2008, with the help of a wonderful team of Puerto Ricans and volunteers from all over the world, the project had become a well-established and respected rainforest research project, sustainably producing tropical hardwoods with minimal disturbance of existing forest ecosystems.

After re-entry, I joined The Planetary Coral Reef Foundation (later to become Biosphere Foundation) and am, to this day, the Foundation's CFO as well as Director of Agriculture and Forestry. (The fact that Gaie, Laser and I have continued to work together for over 30 years, is for me a tribute to our time together at Biosphere 2). During those years, I was also working with the team designing, developing, and constructing the miniature Laboratory Biosphere. My area of responsibility was managing the soils and various crops tested in a series of experiments. I distinctly remember my first entry into this tiny opaque tank lit by high intensity lights to tend the crops. Taking in a deep breath of air and smelling that unique smell that only a closed system can produce, a mixture of soils, plants, microbial life and technics, I was instantly transported back to Biosphere 2.

There have been many other highlights in my time since the Biosphere 2 days. For example, helping run the many Institute of Ecotechnics conferences held annually in France. Participating as camp cook for a month at the Haughton-Mars Project's research station on Devon Island in the Canadian Arctic. When not cooking, I was able to participate in such bizarre activities as helping the geologists search for rocks that might contain microbial life as analogues for Mars exploration, and trying on space suit prototypes being developed for future missions.

The question of how long-term missions in space will deal with communication and conflict resolution has always intrigued me and I have spent much of my time looking into the organizational structures and communication methods of different groups. In my view, the way this is handled will, barring operational catastrophe, make or break any long-term effort to live on another planet. I have not found any perfect solutions, but things seem to go best when the leaders and facilitators are people capable of deep listening. Just as we learned to "listen to our biosphere" in the Biosphere 2 days, we are also going to have to learn to listen to each other.

Eventually my adventures brought me to Bali, Indonesia where The Biosphere Foundation had been establishing community-led conservation projects for some time. Here, I currently split my time between agriculture, forestry projects, and environmental education.

My participation in the design and development of Biosphere 2 and the two years that I spent inside that exquisite mini-world, have been gifts that I treasure. Currently, as Director of Education for Biosphere Foundation, much of my time is spent describing this experience to hundreds of young people from all over the world. The scenario often looks like this: twenty-five sixteen year olds stare in puzzlement at a screen with an old picture of Biosphere 2. The image is not sharp and I hastily explain that the project was designed and built years before the age of digital photography. It was only towards the end of the project that we used an early version of email. No cell phones? No internet? No social media? I can tell from the looks on their faces that I may as well be referring back to some medieval time beyond their comprehension!

I continue to describe and define a biosphere, constantly comparing the strange structure projected in front of them to our planet Earth. What does it mean to be open to energy, open to information, closed to exchange of matter? I define matter with them. Is planet Earth completely closed to matter? At this point generally a few hands shoot up. What about meteors? What about the Mars probes? It is generally agreed that apart from a very small amount of matter, planet Earth is essentially closed to material exchange.

"Open to information" however, is usually a new concept. I talk about the Moon landings and how the communication between the astronauts and Mission Control was information exchange. The Moon landings? For most of these youngsters the Moon landings are also ancient history! Many have never seen the old footage. I describe watching the first landing on the TV, aged 14, and "yes," I assure them, we did at least have TV when I was a teenager!

What comes home to me again and again as I give these talks, is that the Biosphere

2 project was so incredibly ahead of its time, in both its scope and achievements, that it is still capable of capturing the attention and the imagination of nearly every young person I meet. The story unfolds and more and more hands shoot up as the reality of living as part of a closed system begins to sink in. How did we eat, drink, breathe? What animals did we have, what about toilet paper? Where did the toilet flush go? By the end of the talk, the idea that there is no such thing as "away" and that we are living on a planet with finite resources, has taken on a whole new meaning. Often the talk is an introduction to one of Biosphere Foundation's week-long courses in Biosphere Stewardship.

Biosphere 2, and what we learned from the experiment, is an important part of the discussions. In fact, one of Biosphere 2's primary objectives was creating a biosphere on a scale that people could see and understand; something that truly brings home the meaning of what it is to live as part of a fragile, continuously evolving sphere of life. In this respect it succeeded spectacularly, reaching and inspiring millions, and it continues to do so. Over the years I have learned many ways, from the work and example of others, to connect young people to the natural world around them, but my most valuable and impactful tool has always been the Biosphere 2 story.

MARK VAN THILLO (LASER)

Upon the completion of the two-year experiment, and the successful integration of the new technical systems to improve the well-being of Biosphere 2, I joined the PCRF/Biosphere Foundation team as the Chief Operations Officer. Before the Biosphere 2 experiment, I had been the Chief Engineer onboard the *RV Heraclitus*, the Institute of Ecotechnics' ship, and returned after the Biosphere 2 experiment to ensure that this vessel was outfitted and managed with the capacity to operate safely on its new voyage around the world to study and map coral reefs.

I had spent several years traveling through Europe, India, Nepal, Mexico, and central America as a teenager and found that my technical training had given me key tools that were critical for sustaining complex expeditions at sea. Traveling brings the love of exploration, and a life at sea is the closest thing to learning how to voyage on a spaceship—there is nothing like an unbroken horizon and the necessity to navigate and take care of your vessel at all times. Thus, when I heard about the Biosphere 2 project, I realized that this was a precursor to a manned mission to Mars and that all my previous experience would be put to the test. I arrived at the site with my library of Mars books and joined a small group of fellow space enthusiasts to prepare for the two-year experiment as if we were leaving everything behind as we knew it, and so we were. Daily, we would ask ourselves, what do we need to live and not just survive in Biosphere 2 for two years? What does it take to embark on a journey off this planet and live elsewhere? What do we need to bring with us to be happy and healthy on such a voyage?

As a youngster, I loved to take equipment apart and put it back together to understand how it worked. I innately experienced that technical systems are an extension of life, so I saw that the Biosphere 2 technological systems were critical and were intertwined with ecology. Already, my travels had shown me that our future would require us to enhance information exchange between technical and living systems. It became my task to understand every technical system in the Biosphere, which has led to my work still today. At the Biosphere Foundation I have repaired and rebuilt ships and was the technical director of our Wastewater Gardens® systems, Mars On Earth® design, and Laboratory Biosphere experimental chamber.

Video provides the perfect means of conveying complex ideas simply, and my small video studio inside Biosphere 2 ignited a passion to use film as a form of expression. Thus, I initiated 'Studio of the Sea;' an onboard video studio, which was analogous to the one I had personally brought into Biosphere 2 in order to create and record music, as well as edit numerous video clips about what we were witnessing in remote places. 'Studio of the Sea' began in 1995 with the aim to make videos about the beauty of the ocean, its challenges, and its people. We continue to make these simple but powerful educational videos and use them in our educational programs.

Currently, I am the Captain of *Mir*, a lovely classic sailing ship that was constructed in Holland in 1910. She was the first training ship for women sailors and became well known as such in the Mediterranean. During the past decade (2009-2019) with a team of volunteers, we rescued *Mir* in Malta and rebuilt her; it was the biggest recycling project I have ever undertaken. We removed the entire original teak wood and iron deck (it had 2 different deck layers) and reused it for the new teak deck, as well as for all the interior floors. Additionally, all the old equipment, such as a Rolls Royce engine (from 1952), was replaced with state-of-the-art efficient systems to make her a modern-day vessel including solar panels, high-tech masts, rigging and sails.

The Biosphere Foundation is now building a center in North West Bali for education and training about how to care for our biosphere life support system. The center is designed for 32 students and 8 teachers and it will support learning activities that complement our decade-long experience working with the local community in Bali to implement simple, but effective, means of living well with the land and sea. It will also provide a demonstration for sustainable living. Its design incorporates technologies such as: rainwater catchment, solar panels and LED lights to reduce energy usage; no air conditioning, but landscaping of large trees to ensure shade wherever possible, and buildings with high ceilings, fans, and cross ventilation; sustainable building materials; grass block roads and parking areas to allow drainage back into the soil; composting; Wastewater Gardens® for black and grey water; and agriculture, horticulture, and medicinal gardens.

Once the center is complete, we will send *Mir* on her next voyage in the Coral Triangle to bring our local leaders and remote island peoples together to help restore coral reefs. These two bases, one on land and the other at sea, provide the Biosphere Foundation the opportunity to apply what we learned with Biosphere 2, and collaborate with others to inspire a new approach to living well with our Earth's biosphere.

ROY WALFORD

Roy had qualified as a general practitioner to be able to run the medical laboratory in Biosphere 2 and serve as our in-house doctor. He happily returned to his laboratory at the UCLA Medical School where, as professor of pathology, he'd worked for decades developing his revolutionary calorie-restricted diet, as well as working on anti-aging, and life-extension methods.

With his enormous trove of medical and physiological data from Biosphere 2, over the next few years Roy published ten papers on the impacts of the calorie-restricted diet on the biospherians' health, the impact of lowered oxygen despite constant air pressure (unlike an actual mountain climb), and other biospheric health issues. Since the biospherians followed, without fail, the high-nutrient and low calorie diet he helped develop, these medical research papers confirmed that we had the same health results as was predicted: lowered blood cholesterol and blood pressure, and a strengthening of our immune systems. Though the crew initially lost 16% of our body weight, during the second year our weights stabilized, and some weight was regained. Five of the crew had volunteered to spend 24 hours in a metabolic chamber at the VA hospital in Phoenix shortly after our closure experiment was over. This gave further insight into the physiological changes we'd experienced, namely, the data showed that we became more efficient at utilizing the nutrients in our food as a result of the biospheric diet.

The author of over 325 papers in scientific journals and numerous books, Roy published two books after Biosphere 2: *The Anti-Aging Plan* (with his daughter Lisa) and *Beyond the 120-Year Diet*. Roy also published "Biosphere 2: The Voyage of Serendipity" in which he argued that holistic as well as detailed science are crucial to respond to our ecological challenges.

He received numerous awards including ones from the American Society of Clinical Pathology, the American Aging Association, the Gerontological Society of America, and the American Geriatrics Society. Diagnosed with ALS (Lou Gehrig's disease), Roy bravely continued working until his death at age 79 in 2004. After his death, Roy received awards for lifetime achievement from Integrative Medical Therapeutics for Anti-Aging and the American Federation for Aging Research.

Biosphere 2 was monitored by over 1,000 sensors, many giving data every 15 minutes. These systems helped ensure the health of the biomes and humans while providing a wealth of research data.

Research Highlights

"The achievements of the biospherians go beyond the application of state-of-the-art methods of sustainable agriculture. Biosphere 2 recreates, in miniature, the flows and balances that occur on Earth—but it moves through these cycles on 'fast-forward.' Carbon dioxide turnover on Earth takes about three years: in Biosphere 2 it takes about three days. On Earth it takes years or decades to see how changes in the rainforest affect the growth of sorghum or sweet potatoes in another part of the world; in Biosphere 2, the impact may be seen in a matter of weeks. In Biosphere 2, agricultural materials such as crop nutrients and animal wastes recycle through the water and air systems in days, as opposed to weeks or years on Earth. It is, in this sense, an ecological laboratory of incalculable value—the world's largest test-tube."

– **Professor Richard R. Harwood**, C.S. Mott Foundation Chair of Sustainable Agriculture, Michigan State University, September 1993

FOR INTERESTED READERS, we have included some representative graphs, tables, maps, and a sample of the detailed monitoring that was carried out in Biosphere 2 to illustrate the different kinds of challenges we faced.

CARBON DIOXIDE AND ATMOSPHERIC MANAGEMENT

One of the wonders of Biosphere 2 was the daily and seasonal variations in CO_2, which were much larger than on Earth. This led to the practice of intervening as stewards to manage the atmosphere and ensure there was not a runaway rise in CO_2. The information and data gained provides relevant knowledge about how we can maintain CO_2 and other greenhouse gases today in order to minimize the severity of climate change. We learned rapidly that we could maintain a healthy atmosphere by maximizing plant biomass and productivity. Any unused space was quickly turned into gardens, and vegetation (plants or algae) was pruned to spur new growth. Additionally, we created carbon sinks by storing biomass, such as stockpiling and drying savannah grasses, and minimizing seasonal activities that produced more CO_2, such as composting, worm beds, and tilling soil.

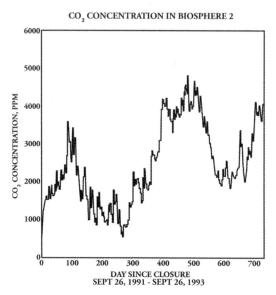

Carbon dioxide dynamics during the two-year closure experiment. There was strong seasonality; in the longer sunlight months, concentrations declined, reaching a low of below 1000 ppm during the 1991 first summer. In both winters, CO_2 concentrations rose, reaching near 5000 ppm during the second winter of 1992-1993.

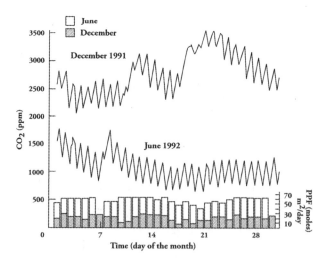

This graph shows Biosphere 2's large daily changes in carbon dioxide. The days from December 1991 and June 1992 also illustrate the tight correlation between the amount of sunlight and the levels of atmospheric CO_2. CO_2 levels were between 2000 ppm and 3500 ppm in December 1991, while they were from 1000 ppm to around 1800 ppm in June 1992.

CORAL REEF OCEAN

Maintaining the health of the coral reef ocean was particularly challenging because the high levels of atmospheric CO_2 in Biosphere 2 was absorbed in its waters, lowering its pH levels. Because Biosphere 2 was a closed ecological system, we could monitor minute changes of CO_2 and pH, thus observing dynamics that were magnified, making every day virtually another experiment.

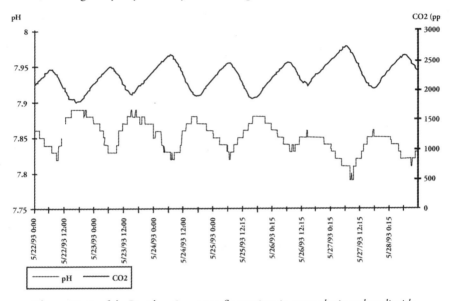

The sensitivity of the Biosphere 2 ocean to fluctuations in atmospheric carbon dioxide levels is shown in this graph. As CO_2 levels fall during the day, due to the uptake of CO_2 through algae photosynthesis, ocean pH increases; as CO_2 levels rise during the night, since system respiration now dominates, ocean pH levels drop.

"Microbial diversity [of Biosphere 2] was rich and varied, comparable to those found in natural environments…One of the beauties of this system is that organisms that are rare in natural environments can become abundant. This closed system is an unbelievable opportunity for microbiologists, a phenomenal laboratory and a real opportunity to do something that can't be done anywhere else."

– **Professor Donald Spoon,** Georgetown University

THE WATER CYCLE

In a closed ecological system with a diversity of biomes, a lot of ingenuity and intelligent management was required to ensure that each biome/ecosystem received the correct amounts and quality of water.

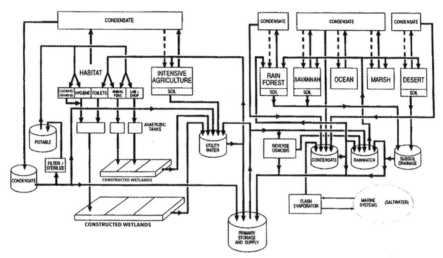

Biosphere 2 Water Systems

This schematic shows the complexity of the water cycle in the Biosphere 2 facility. Innovative technologies, such as our constructed wetlands, that were designed to treat and then recycle all human sewage and other wastewater back to our farm irrigation, were created. Condensation collection supplied potable water and was added to irrigation/ rainwater to minimize buildup of salts in our terrestrial biomes. A flash evaporator was also included as part of Biosphere 2's water cycle to manage salt levels.

CREATING AND STUDYING MINI-BIOMES

Prior to Biosphere 2, it was unprecedented to create a laboratory with such a wide diversity of land and aquatic/marine biomes; from rainforest to desert, freshwater marsh to a coral reef ocean, to agriculture/farm and a human habitat in one facility. In addition, to further mimic the mosaic of natural biomes, each area had internal zones dominated by characteristic species. Biosphere 2 contained the first and only rainforest and mangroves in the state of Arizona. Given the current degradation of Earth's natural biomes, the creation of these analogue biomes offered a unique opportunity to learn more about these biomes that are so vital for Earth's biosphere.

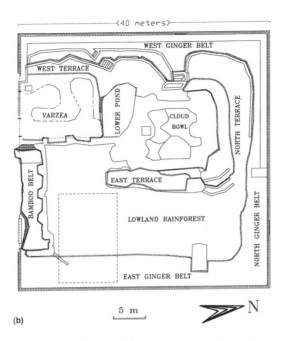

This map depicts the ecological zones of the Biosphere 2 rainforest. The planned succession used fast-growing first-generation trees to protect the more sunlight-sensitive trees which would supplant them. The perimeter "ginger belt" prevented harsh side light from damaging the core rainforest. A bamboo belt in between rainforest and ocean was designed to trap salt aerosols and prevent salt damage.

"Restoration ecologists understand the value of the process of ecosystem construction as a means of evaluating hypotheses and gaining new information about ecosystem function. The synthetic approach may raise more questions than it answers, but new questions and new ways of thinking about ecosystems and the biosphere are critical to the advancement of our knowledge. In our view, the ongoing process of creating and maintaining ecological systems within Biosphere 2 will continue to reveal important information about the Earth's ecosystems for many years to come. Some of this information may be directly applicable to the practice of restoring these ecosystems."

– **John Petersen, Alan E. Haberstock, Thomas G. Siccama, Kristina A. Vogt, Daniel J. Vogt and Barbara L. Tusting,** Yale University School of Forestry and Environmental Studies

Terrestrial Biomes

In order to document and track the evolution of biomes, every plant was mapped and measured both before the closure experiments began and afterwards. Over eleven thousand individual grasses, shrubs and trees were inventoried; a dataset that showed which ones thrived, survived, or were lost while sensors kept a detailed record of environmental conditions. Since Biosphere 2 had elevated CO_2, this could have provided valuable insights into the ecological consequences of climate change, but the data was apparently lost when the project changed hands in 1994.

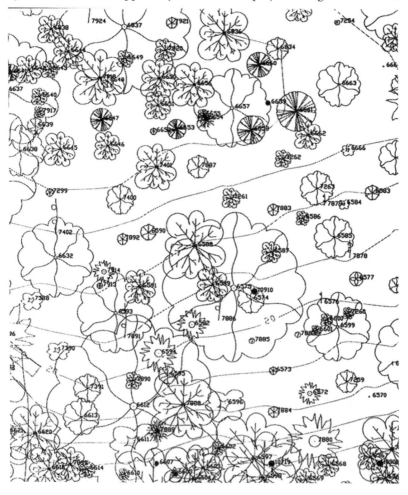

This map of the rainforest done prior to closure. All of the Biosphere 2 terrestrial biomes, marsh and ocean were mapped before and after the two-year experiment.

Everglades Marsh Estuary

Many miles of a natural Florida wetland system were compressed into a compact Biosphere 2 marsh biome. On the inlet side was a freshwater zone that was followed by five progressively saltier wetlands, each with its own dominant vegetation. The marsh/mangrove grew rapidly, proving to be an excellent carbon sink and successfully replicating the ecology of the Everglades areas where it was collected.

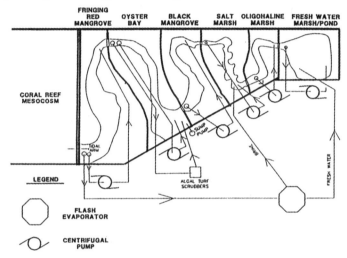

Ecological zones of the marsh/mangrove system modeled on a tropical estuary which was collected in the Florida Everglades. An ecological transect runs from the least salty inland freshwater marsh through varying ecosystem salinities until reaching the red mangroves.

"Examination of the ecosystem within Biosphere 2 [showed] features of self-organization...it appeared to be reinforcing the species that collect more energy...species diversity of plants was approaching normal biodiversity...the observed trend (carbon dioxide absorption by carbonates and high net production of "weed species vegetation") if allowed to continue... would eventually generate enough gross production to match respiration of the soil, which was gradually declining. Thus, the self- organizational development of a human life support was successfully underway...the smaller, faster Biosphere 2 is a good model for studying the biogeochemical dynamics of our earth."

– Professor Howard T. Odum, University of Florida, 1996

A MARVEL OF ENGINEERING

Biosphere 2 was nearly airtight with an air exchange under 10% per year. It was orders of magnitude more sealed than any previous closed ecological system or spacecraft, despite its size and exposure to freezing winters and very hot summers, with miles of seals on its extensive space frame roofs and on the stainless steel liner underground.

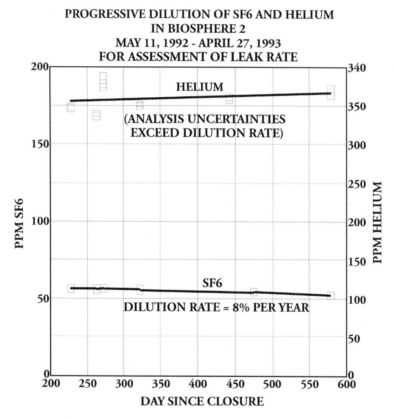

The measured airtightness of Biosphere 2 was determined by monitoring the decline of the inert gas sulfur hexafluoride (SF6), which was injected into Biosphere 2. This trace gas showed us that air leakage was around 8% per year.

The airtightness allowed us to track small and long-term changes in our atmosphere such as the decline in atmospheric oxygen. It also allowed for detailed monitoring of a wide variety of trace gases such as outgassing from technologies and man-made materials or those released from soils and living organisms.

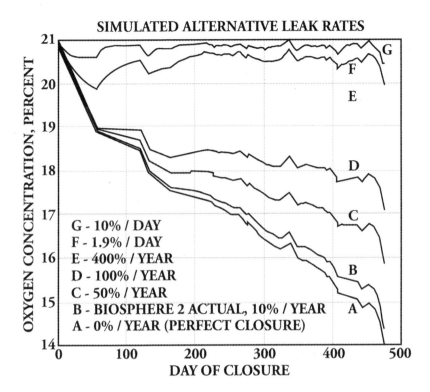

This graph represents how a gas like oxygen, which declined in Biosphere 2, could be monitored. Line A is extrapolated to see what the oxygen decline would have been with a 0% leak rate; line B were the measurements that were taken with a 10% leak rate. The other lines represent what would have been measured if our leak rate had been higher. Had our leak rates been as high as those of Lines E, F, and G, we would not have realized oxygen levels were declining in our atmosphere.

We believe results obtained in this experiment in combination with global studies of Earth biosphere through remote monitoring and extensive application of scientific technology will not only contribute to a better life on the Earth but will also serve the task of the human planetary exploration of the solar system."

– **Academician Oleg Gazenko**, Director of the Institute of Biomedical Problems, Russian Academy of Sciences, Moscow

CYCLES OF AIR AND WATER

Every mini-world will have different ratios of soil, plants, water, and atmosphere. This was one reason we hoped Biosphere 2 would prove to be the beginning of a science that could compare biospheres, and thus provide a basis for the study of biospherics to provide insight into how our Earth biosphere functions.

	EARTH	BIOSPHERE 2
Ratio of biomass carbon: atmospheric carbon	1:1 (at 350 ppm CO_2)	100:1 (at 1500 ppm CO_2)
Ratio of soil carbon: atmospheric carbon	2:1	5000:1
Estimated carbon cycling time (residence time in atmosphere)	3 years	1-4 days

The abundance and concentration of life with a small atmosphere inside Biosphere 2 caused a large acceleration of the movement of carbon dioxide into and out of the atmosphere.

Reservoir	Volume (liters)	Percentage of Biosphere 2 water	Typical daily water flux (10^3 liters)	Estimated residence time Biosphere 2	Estimated residence Earth biosphere	Acceleration in Biosphere 2
Ocean/ Marsh	4×10^6	~61%	3.4	~1200 days	3000-3200 years	1000 times faster
Soils	$1-2 \times 10^6$	~23% (calculated on 1.5×10^6)	Irrigation: 18.3 Soil Drainage: 4.6 -Plant uptake and ET	~60 days	30-60 days	similar
Atmosphere	2×10^3	~0.03%	12.4 (evapotranspiration)	1-4 hours	9 days	50-200 times faster

Size of water reservoirs and water cycling times in Biosphere 2 and Earth's biosphere. Biosphere 2 atmospheric water cycles were 50-200 times faster and water in the small ocean and marsh cycles one thousand times faster.

DIET FOR A SMALL WORLD

Biosphere 2's farm was one of the most productive half acres in the world. It supplied over 80% of the total diet for the first two-year closure. After system improvements, including selecting of crops better suited for Biosphere 2's conditions, the farm supplied all of the second closure team's diet.

CROP	TOTAL 2 YR YIELD KG	GRAMS PER PERSON PER DAY	PROTEIN (g.) PERSON	FAT (g.)/ PERSON	Kcal/ PERSON
GRAINS: RICE	277	47	4	0.9	168
SORGHUM	190	32	4	0.6	107
WHEAT	192	32	4	0.7	108
STARCHY VEGETABLES: POTATO	240	41	1	0.6	31
SWEET POTATO	2765	468	7	1.3	494
MALANGA, YAM	2	20	12	0	22
LEGUMES: PEANUT, SOY BEAN, LABLAB, PEA, SOYBEAN, PINTO BEAN	208	60	13	13.2	269
VEGETABLES: BEET GREENS, SWEET POTATO GREENS, CHARD	637	108	1	0.2	22
BEET ROOTS	760	129	2	0.4	57
BELL PEPPER, GREEN BEANS, CHILI, CUCUMBER, KALE, PAK CHOI, PEA	331	57	1	1	15
CARROTS	225	38	0	0.1	17
CABBAGE	153	26	0	0	6
EGGPLANT	245	41	0	0	11
LETTUCE, ONION	289	49	0	0.1	11
SUMMER SQUASH	513	87	1	0.1	17
TOMATO	353	60	1	0.1	12
WINTER SQUASH	343	58	1	0.2	37
FRUITS: BANANA	2171	367	2	10.5	220
PAPAYA	1216	206	1	0.2	53
FIG, GUAVA, KUMQUAT, LEMON LIME ORANGE	133	23	0	0.1	11
ANIMAL PRODUCTS: GOAT MILK	842	142	5	5.6	99
GOAT, PORK, FISH, CHICKEN MEAT, EGGS	94	108	3	3.1	38
TOTAL PRODUCED	12432	2107	53	39	1823

Food production from the half acre Biosphere 2 farm during the 2-year experiment. The diet was primarily vegetarian with some tropical fruits and small amounts of eggs, milk, and meats. Over the course of the 2 years, the biospherians became healthier on our calorie-restricted but nutrient-dense diet and successfully integrated working together in the farm to produce food daily.

BIOSPHERE 2'S ENERGY SYSTEM

Like the Earth's biosphere, our plants were dependent on sunlight for their growth and they grew better in longer daylight seasons. The Biosphere 2 energy center had to supply energy for all vital technologies inside which supported that life. A loss of cooling during the summer months, even for a few hours, would have been catastrophic, so backup systems and redundancy were vital.

Energy	Peak	Average
Electrical	500 KW	200 KW
External Support	2000 KW	1200 KW
Heating	11×10^6 kj/hr	5×10^6 kj/hr
Cooling	30×10^6 kj/hr	10×10^6 kj/hr
Solar flux	27×10^6 kj/hr	6×10^6 kj/hr
Photosynthetically Active Radiation (PAR)	30 moles m^{-2} day^{-1}	moles m^{-2} day^{-1}

Biosphere 2 energy flows. Cooling was the largest demand on our external energy center. Photosynthetically active radiation from sunlight powered the growth of green plants.

"Has a time of experimentation with large-scale Biospheres come? The tradition of using small-scale microcosms and growth chambers does not capture the essence of whole system responses, a scale that will affect humanity. Biosphere 2 will continue to stimulate the minds of those who have the vision to think beyond the veil of tradition. As much as anything else this technology, or conglomerate of them, may play a vital role in the emergence of new sciences due simply to the fact that this tool enables experimental work at a scale that rarely has been possible."

– **Professor Dr. Bruno D.V. Marino**, Columbia University, and **Professor Howard T. Odum**, University of Florida

NOTE FROM THE PUBLISHER

I WAS ONE OF THE LUCKY PEOPLE involved with the design, building, and operation of the Biosphere 2 project from 1984-1994. Synergetic Press began publishing books on biospherics in the early 80s starting with *The Biosphere Catalogue*, followed by the first translation of Vladimir Vernadsky's *The Biosphere* in 1986. Biosphere 2 was designed, in part, to put Vernadsky's theory to the test: was there such a thing as a biosphere, a self-sustaining, co-evolutionary sphere of life on Planet Earth?

During construction of Biosphere 2, the Visitor Center was built and accompanying educational programs and classroom curriculum were developed. This was before the internet existed, so, in order to reach millions of students around the world, we hosted numerous satellite uplinks with the biospherians who spoke with classrooms and schools all over the world from inside the experiment. We were committed to sharing the lessons we were learning as we strove to build a miniature world.

This particular book was written by the crew while they lived inside the facility. They sent their work out in sections via Biosphere 2's on-site email system—known as Dialcom—the world's first commercial email service. I would take typeset printouts of the book to hold up at the window in order to show the authors. The first edition of *Life Under Glass* was published the day they came out in September of 1993, an event viewed on television by millions of people around the world.

In addition to our current managing editor, Amanda Müller, I'd like to thank my colleagues who worked on the first edition at the Biosphere Press: Linnea Gentry, Kathy Horton, Lynn Ratener, and Anthony G. Blake. Special thanks to the fine photographers whose photos appear in this book: Peter Menzel who captured the taste of history and adventure in his images; and other photographers I worked with at Biosphere 2 over a period of years, Gill C. Kenny, Gonzalo Arcila, C. Allan Morgan, and Mike Stokolos. Special credit goes to crew member, Abigail Alling, who passionately captured their two years living under glass from behind the lens, as well as carrying out all her many other duties.

May the critical lessons we learned about planetary stewardship pass on to others through the retelling of this story.

– **Deborah Parrish Snyder**
Publisher, Synergetic Press
Santa Fe, New Mexico, March 2020

INDEX